THE WORLD WE ARE MAKING

A Wealth of Trees

ALAN C. JENKINS

A Wealth of Trees

Forestry and the Use of Timber

ILLUSTRATED WITH PHOTOGRAPHS AND DIAGRAMS

Methuen & Co Ltd
II NEW FETTER LANE · LONDON EC4

Also by Alan C. Jenkins
THE GOLDEN BAND
THE SILVER HAUL

First published in Great Britain 1969

Phototypeset by BAS Printers Ltd, Wallop, Hampshire
and printed in Great Britain
by Ebenezer Baylis & Son Limited
The Trinity Press, Worcester

SBN 416 44840 2

Contents

List of Illustrations

PHOTOGRAPHS

LINE DRAWINGS AND DIAGRAMS

Acknowledgments

In writing this book I have referred to the following sources: *Forests and Seapower* by R. G. Albion; *Défense et Illustration de la Forêt* by J. B. Bavier and A. Bourquin; *Treasure of the Timberlands* by Jim Breetveld; *A Tour through the Whole Island of Britain* by Daniel Defoe; *Trees, Woods and Man* by H. L. Edlin; *Sylva* by John Evelyn; *Landscape with Trees* by Miles Hadfield; *Holy Old Mackinaw* by Steward H. Holbrook; *Story of the Saw* by P. d'A. Jones and E. N. Simons; *Wood from Forest to Man* by Bryan Latham; *British Sawmilling Practice* by Victor Serry; *The Wheelwright's Shop* by George Sturt.

For information or material placed at my disposal I am grateful to the Commonwealth Forestry Institute, the Forest Products Research Laboratory, the Canadian Department of Forestry, the U.S. Forest Products Laboratory, the New Zealand Forestry Service, the East Malling Research Institute; also to the Schauman Company of Jyväskylä in Finland for visits to their factories, and to the Conservator of the Southwestern Region of the Forestry Commission for visits to plantations in his area. Lieutenant-Colonel D. B. Rooke M.C. of the Forestry Commission, kindly read the typescript of the book.

Author and publishers gratefully acknowledge help received, and permission to reproduce illustration material, from the following:

Australian News and Information Bureau (photos nos. 22 and 45); The High Commissioner for Canada (photos nos. 11, 25, 26, 28, 42); Bowater Paper Corporation (photos nos. 39 to 41); Forestry Commission (photos nos. 2 to 7, 9, 12 to 15); Food & Agriculture Organization of the United Nations (photos nos. 17 and 21); The High Commissioner for New Zealand (photos nos. 19, 23, 30, 31, 34, 35); Novosti Press Agency (photos nos. 1 and 33); Radio Times Hulton Picture Library (photo no. 18 and Figs. 1 to 7); Messrs Stenner of Tiverton (photos nos. 36 to 38); Timber Research & Development Association (photo no. 44 and permission to reproduce Figs. 5 and 6); United States Information Service (photos nos. 8, 10, 16, 27); Weyerhaeuser Company of Washington (photos nos. 20, 24, 29, 32).

1 · The Quest for Timber

Against the dense jungle trees that rose in a green chasm, the roar of the hovercraft echoed frenziedly as the vessel skimmed along the turbid river. The men inside the comfortable saloon gazed out with mild interest at what to them was a familiar landscape.

These men were neither tourists nor explorers. They were commuting to their place of work far in the jungle after a spell of leave down river. They were foresters: specialists in many different branches of their profession – lumbermen, scientists, botanists, entomologists. In this hovercraft they could ride into the heart of a once-impenetrable jungle. Soon they would be putting aside their cameras and fountain-pens and writing-pads and exchanging them for gas-saws, dendrometers, humidity-sticks, intricate graphs.

They knew they were approaching their destination, for together with the steady throb of the hovercraft's engines, another noise mingled. Above the jungle rose a giant helicopter, its cargo of tree trunks dangling beneath it. In the distance, too, lumbered an enormous balloon, manœuvring slowly as it extracted a massive tree trunk from the green phalanx that stretched out seemingly without end.

1 *A balloon being used in Russia to extract timber*

That is no fanciful picture but a distinct possibility of the shape of things to come in forestry. In spite of all the concrete, steel, plastic and other substitutes for wood and wood products, mankind is still vastly dependent on timber in all its forms. Two thousand million cubic metres of timber are taken from the forests of the world every year. Yet even that immense quantity is nothing like enough for our needs. It is estimated that by 1975 fifty per cent more wood and wood products will be required than eight or ten years ago.

Not only is the world population exploding, but also the standard of living in the developing countries is rising rapidly: people require better accommodation, better communications. With the increase in literacy, millions more people are able to read and their demand for more books and newspapers and journals becomes almost as insistent as their demand for food.

All this means an ever-increasing need for wood. In every country, developed and developing, urgent efforts are being made to cope with this need. New forests are being created with scientific precision, new ways of improving tree-growth are being evolved, new methods of utilizing wood are being discovered. At the same time, the hitherto almost untouched forests of tropical countries, like South America, are being opened up because of advances both in transport and in medicine.

The nineteenth-century American author, Henry David Thoreau, wrote that wood to the settlers of his country had a more permanent value than that of gold. The Finns and Swedes, too, refer to their timber as green gold because of its importance in their economy. It is the purpose of this book to provide a glimpse of Man's age-old exploitation of and dependence on this natural treasure, and to show something of the work entailed in modern forestry and wood-usage and the developments taking place.

2 · In the Past

Men have always been fascinated by trees. Long before trees began to play a vital part in human economy, when food was his chief consideration, Primitive Man went in awe of the trees and forests.

In Europe forests covered the greater part of the land. What human settlements existed were like moth-holes eaten in a vast, smothering green quilt.

As recently as two thousand years ago, much of Britain was covered by oak forests. Place-names derived from the ancient words for the oak are common – Aikton, Darach, Derry, Derwent, Eggington, Eyke, Okeford and so on.

In what was to become Germany it was said that men (if they were bold enough) could travel through the Hercynian forests for two months without reaching open country. Northern Italy was lost in almost impenetrable forests of oak and chestnut. Even in the days of the Gauls, two-thirds of France was forestland.

Fig. 1 *The oak worship of the ancient Britons*

These forests were a threat to Early Man. Not only did he fear the wild beasts, but he regarded the forests as being the haunt of terrible demons whom it would be fatal to offend. As a result of such an attitude, the worship of trees was a frequent practice. Among the Celtic Druids, oak-worship was part of their religion, and sacred groves of trees were in fact the earliest form of temple. As well as in Britain, these groves existed in ancient Germany, Sweden, and along the Volga where Stalingrad now stands. Dire punishment, such as the loss of a hand, was dealt out to anyone rash enough to harm a tree dedicated to the gods.

Even today, Slavic and Irish people in very remote places still adorn certain trees with articles of clothing in return for favours granted.

Some American Indians regularly consulted the cottonwood tree which they were convinced was so intelligent it would advise them in their schemes. Even when a tree had to be cut down, forgiveness was asked or penance done in countries as far apart as Austria and Borneo.

But though many other folk believed spirits inhabited trees, they did not always treat them with the same respect. In Japan, fruit-trees were ceremonially asked if they would yield a good crop the following year; otherwise, it was threatened, they would be cut down!

All over the world, wherever trees grew, this superstitious attitude prevailed.

In times when nature was an even greater mystery than it is nowadays, it is not surprising that men should be impressed by the mighty image of the tree. But far more vital was the material role it played in daily life.

As a baby, Man was rocked in a wooden cradle (people such as the Lapps still carry their infants in cradles fashioned from a birchlog). At the end of his life, he was laid to rest in a wooden coffin.

In between, there were few activities man could indulge in without the assistance of the tree and all its products. Home and farming, hunting and fishing and soldiering were all dependent in one way or another on the use of wood.

The turf hut of the Saxon had a framework of wooden poles, while his 'burh' or settlement was protected by a wooden stockade. Many centuries later, American pioneer settlements and forts were protected in the same way against attacks by Red Indians. Today, more timber houses exist than dwellings made of brick or stone in countries such as North America, Scandinavia and Russia.

As dwellings developed from elementary hovels, they became more dependent on timber, from the appropriately named tree-nails of wood to the equally well-named roof-tree, the main beam of which sometimes consisted of a whole straight, trimmed tree. As these dwellings became homes rather than merely shelters, they were decorated with handsome wood panelling, and filled with furniture, all made from the tree.

The stone-built castles, that succeeded the more modest motte-and-bailey type of fortification in Europe, still required sturdy timber for rafters and floor-beams. The massive size of these can be guessed at from the sockets in the walls of many ruins.

In the same way, the drawbridge and portcullis were mainly constructed of timber. The handles of the lances knights fought with, the saddles of the great chargers they rode, the quarter-staffs the men-at-arms sported with, were all of wood.

On the land, wood provided the peasant's plough, and the cart or sled on which he hauled home the logs needed for warmth and cooking. The hurdle-fences, which were then more typical than bank or hedgerow, were made from coppice-shoots grown from the stumps of felled trees.

Ox-yokes, mill-wheels, spinning-wheels, clothes-chests, musical instruments, tallies, pails, and the many household articles significantly known as treen, and which included bowls, platters and so forth – it would be easier to list what was not made of wood than all the countless articles that were.

Even water-mains were made of bored tree trunks. As recently as 1900, four hundred miles were still in use in London and the tree trunks were bored by a horse-mill situated near Fleet Street. Footwear was often either made from a wood-product or carved from wood itself, like the Breton sabots. Rawhide was processed into workable leather with the aid of tannin obtained from trees,

Fig. 2 *Wooden household implements*

while the shoes Russian moujiks wore came from bast, the inner bark of trees. In times of starvation this inner bark was used to eke out flour, as for example in the famine that scourged Finland in the last century. A less serious shortage, that of tobacco, was sometimes overcome by American Indians through the use of kinikinic, the inner bark of the red willow. For centuries primitive tribes have obtained their alcohol from the fermented sap of various trees.

Quite apart from genuine fruit-trees that have gradually been developed, the fruit of wild trees has been an important item in human economy, both for men and their animals. For instance, pigs were largely nourished on the mast or fruit of oak and beech, and bacon was cured with the aid of another by-product of the tree: the sawdust from the sawing-pits. Nearly every part of the tree had its use.

3 · Industry Cranks up

Of course Man had to avail himself of everything that Nature offered. Things went wrong only when he became thoughtlessly greedy.

At first his demands on the trees and the forests were modest enough. But as Man progressed and his standard of living improved, his demands on natural resources were increasing rapidly all the time. Among the chief resources which he plundered were the forests. They were the most easily accessible and timber was the most easily processed.

Forest clearance had to some degree been going on for thousands of years – ever since men ceased to live entirely by hunting and fishing and began to turn, however primitively, to a form of agriculture which required a more settled existence.

In Britain it was the Neolithic or New Stone Age people who were the first, with axe and fire, to attack the forests and make clearings for their simple crops and for their animals. They may have tackled first the downlands of south-eastern England, where the beech trees were less dense than the oak-forests of the valleys. Once they established their flocks of sheep or cattle, these themselves, by their grazing, prevented any come-back the trees could have made.

After the New Stone Age came the Bronze and Iron Ages, each of which in turn led to immense advances in human techniques. The land was ploughed for the first time, instead of being scratched by hoes; more timber was felled by the greatly improved tools – and more was needed to smelt the iron from which such tools were fashioned.

But though a steady amount of forest clearance was carried out once Man became more settled, it was much later that the face of the land began to change with devastating swiftness, through the wholesale slaughter of trees which nobody saw the need to replace.

In England, the reign of Queen Elizabeth I saw the initial stirrings of modern commerce and industry which were made possible by a sudden development of mining of different kinds such as lead, copper, tin, iron ore and coal.

Coal, which eventually took the place of wood as fuel for industrial purposes, is formed from the fossilized trees of an ancient world, which fell through natural causes, long before Man and his axe appeared. But the other forms of mining, such as iron ore and lead, caused the forest to be bled to death. Enormous quantities of timber were needed to make charcoal for the process of smelting the raw material.

Older still than these mining activities and equally inimical to the forests was the making of salt. Great quantities of this were called for in preserving meat, in days when many farm animals had to be slaughtered at the approach of winter through lack of fodder. Nowadays commercial salt is mined underground, while

Fig. 3 *Early industry: wood is being used for fuel in the making of lead ore*

table salt is usually obtained from brine. But in the past it was largely obtained from the evaporation of sea-water in lead-pans which necessitated constant supplies of wood for the fires used in the process.

Glass-making was another trade which demanded large supplies of fuel. When the local supply was exhausted, the manufacturers moved off to some other neighbourhood with a handy supply of timber.

Even the tanning of leather, in never-ending demand for footwear and clothes and harness, required large amounts of oak-bark from which the tannic acid was extracted. Brewing, too, consumed trees wholesale, for hops required tall poles to grow on and the oasthouses for drying the hops were heated by charcoal-burning kilns.

Although all these were peaceful trades, they made ruinous inroads on the forests and woodlands of Britain and other European countries. But in the fourteenth century, a new trend in warfare inaugurated another serious threat to the trees: for in order to propel the cannon balls, even more charcoal was needed, to mix with saltpetre and sulphur in the manufacture of gunpowder.

The charcoal-burner became an almost legendary figure, a stock character in folk-tales and romances, pursuing his grimy calling in remote woodlands and employing methods still practised occasionally in the twentieth century.

This age-old method consists basically of stacking suitable wood – alder, for example – round a central pole which is removed when the pile is complete, so that a 'chimney' is left inside it. The subsequent burning is carefully regulated by covering the stack with soil or turves, while if the wind is too strong wattle-screens are erected on the windward side.

2 *Iron kilns used by English charcoal burners today*

Modern charcoal-burning is carried out chiefly by the 'Stafford' method. In this, small pieces of wood are fed into huge cylindrical retorts, already heated to 300 degrees Fahrenheit. Nowadays wood charcoal is used mainly in various metallurgical and chemical processes as well as for cooking.

4 · Wooden Ships

Among the most rapacious consumers of timber, down to the arrival of iron ships in the last century, were the shipwrights. From the war-galleys of the Carthaginians, the Roman triremes, the longships of the Vikings and the galleons of Columbus, all these ships were products of the tree.

As ships and navies became progressively larger and particularly as war between England and France intensified, an adequate supply of timber created an acute problem. Even as early as the twelfth and thirteenth centuries, England was importing Norwegian timber, and in the days of the Spanish Armada, timber was of such potential value to an enemy that it was regarded as contraband. All the maritime countries, such as Spain, France, Holland, had timber problems of varying degree, but the worst affected was England.

The matter reached a climax during the eighteenth century and the Napoleonic Wars. Remembering David Garrick's patriotic jingle,

> *'Heart of oak are our ships,*
> *Heart of oak are our men,'*

one might imagine that a fighting ship was built of oak throughout.

In fact, the timbers of such a vessel were by way of being international in character. The planking for her sides would have been purchased on the shores of the Baltic. It was likely that the huge mainmast came from the forests of New England or Newfoundland. Her topmasts had probably been grown in Catherine the Great's Russia, while her bowsprit and lesser spars might have come from Norway.

Britain was unable to supply her own mast timber, not because in this case her forests had been cut down, but because the native Scots pine was not sufficiently strong or resilient – or big enough. The mainmast of a 'First-rate' of 120 guns had to be forty inches in diameter and forty yards long, and this was but one of the many masts, yards and booms she required to carry her full canvas of some thirty different sails.

The need for a steady supply of mast timber influenced English foreign policy for centuries. From the days of Cromwell, England despatched more than a dozen naval expeditions to the Baltic because of this.

Long before the War of Independence, the American Colonists had resented British Government inspectors requisitioning their best timber. The official broad arrow became a hated symbol, and more than one government official was thrown into the local pond when he came in search of suitable trees. But after 1775 England was completely cut off from American supplies and Baltic timber

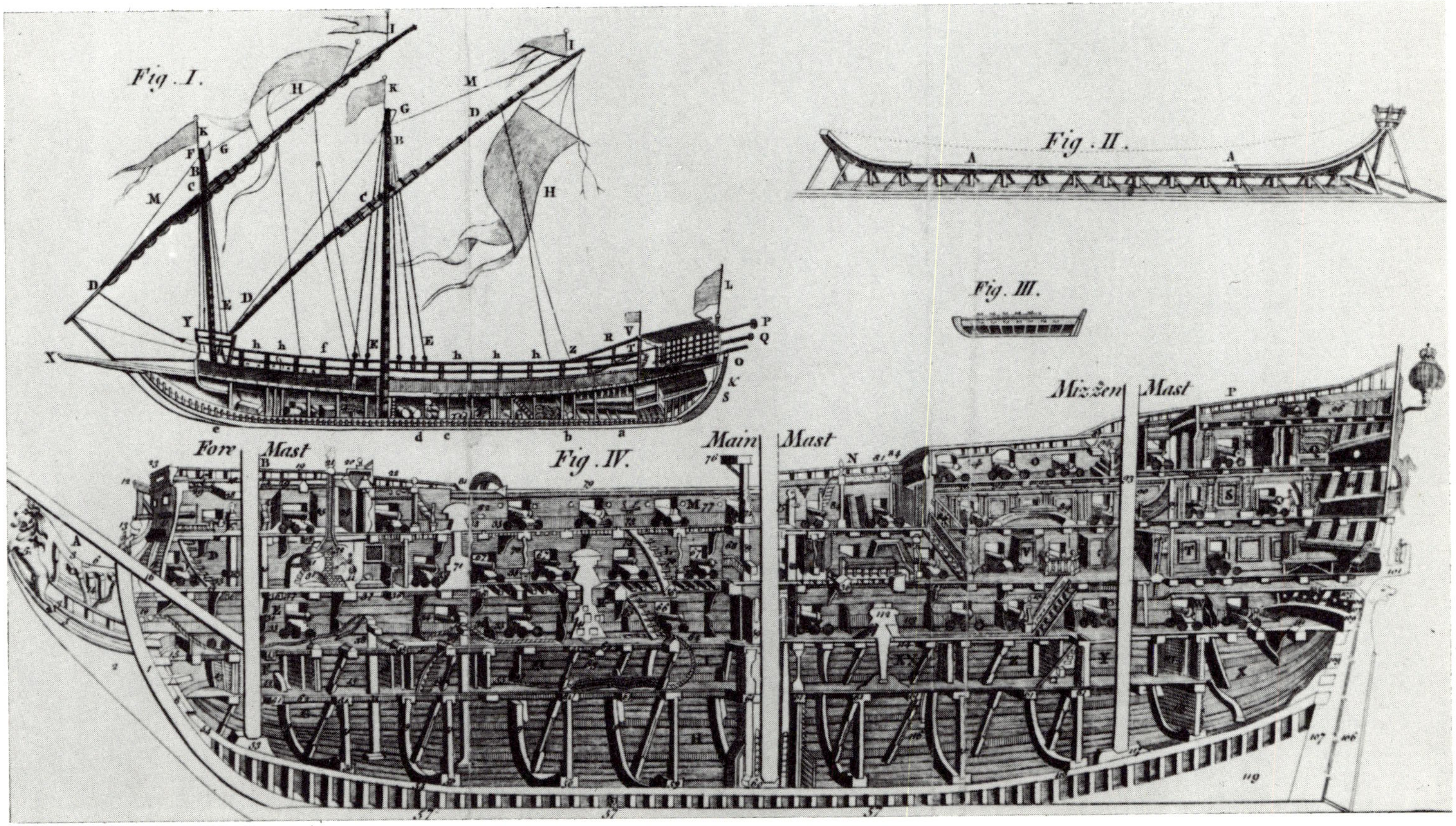

Fig. 4 *Cross section of a 'First Rate Ship of War', showing the amount of timber used in shipbuilding during the eighteenth and nineteenth centuries*

became even more vital, for Canadian resources were astonishingly neglected until later on.

As far as hull-timber was concerned, the best was undoubtedly British. Elm and beech were sometimes used for keels, but were not entirely satisfactory. Vessels were sometimes built entirely of fir, but this splintered dangerously under bombardment.

Oak was by far the most suitable. It was said to be 'tough, bending well, strong, and not too heavy, nor easily admitting water'. The trouble was that it took between 150 and 200 years for an oaktree to grow big enough and nobody had had sufficient foresight to plan a completely adequate supply.

Yet a ship of the line of Nelson's days required the product of two thousand oaktrees for her timbers, which meant cutting sixty acres of woodland planted perhaps in the reign of Elizabeth or James I.

Nor did the matter end there. Ships suffered seriously from dry rot and within a few years of being launched would need extensive repairs. For example, the *Victory* cost some £63,000 when she was built in 1765. By 1815, the total spent on her amounted to more than £370,000, a vast sum in those days.

Even allowing for all the imported timber, it was no wonder the woods of Britain were tumbling down. Not even woods full of trees similar to the famous Newlands oak in the Forest of Dean, with a girth of forty-four feet, could have satisfied the demand.

5 · A Belated Policy

Over the centuries, random, unorganised efforts had been made to preserve Britain's forests. Even William the Conqueror, though selfishly motivated by his love of hunting, could be said to have helped preserve the New Forest by declaring it a royal domain.

Henry VIII decreed that when oak woods were felled, enough young trees should be left for natural regeneration. This, however, was a rather belated act for he had already contributed to the destruction of those woodlands when he sold up monastic property during the Dissolution. The new owners promptly sold the timber for quick profits.

During the seventeenth century, a few enlightened people became worried about the situation. Apart from the maritime and industrial demands, increased quantities of timber were needed for building purposes owing to the population explosion that took place under the early Stuarts. When, later in the century, the Fire of London occurred, the need was made immensely greater.

In 1613, Arthur Standish had published an essay on increasing home-grown timber. Samuel Pepys, Secretary of the Navy in the reign of Charles II, gave warnings of the shortage and certain appropriate measures were taken in the royal forests as a result. One of the greatest advocates of afforestation was Pepys's friend and contemporary, John Evelyn, who published his still-read 'Sylva, a discourse of forest trees, for the encouragement of an Industry and Worthy Labour, much in our days neglected'.

But broadly speaking, apart from a handful of enterprising landowners, like the Duke of Atholl who a century later planted fourteen million larch trees in Scotland, little heed was paid to Evelyn's warning that the country's only protection was its 'wooden walls'. Unless, of course, we count a devoted individual such as Admiral Collingwood, Nelson's second-in-command, who spent his leave planting acorns on his Northumberland estate. During the nineteenth century, with the increased use of coal, the improved methods of making iron and steel, and the replacement of wooden ships by iron hulls, the question of home-grown timber became even more neglected.

Throughout the Industrial Revolution, increasingly large quantities of timber were required for a new range of purposes, such as pit-props, railway sleepers, factory buildings and so on. But it was found cheaper for the timber to be imported.

In France, on the other hand, the Ecole Nationale des Eaux et Forêts was founded in 1824 and the French forestry service has such a proud tradition that its officers are allowed to continue wearing their official forestry uniform when called up to do their military training.

Britain did not fail entirely in this respect. Forty years later, a comparably efficient forestry service was founded. But in India, not in Britain. In India the first Inspector-General of Forests was appointed in 1863 and modern India and Pakistan have maintained the traditions of the service. In the 1880's, a School of Forestry was established in Surrey – but again, not in the interests of home forests. It was to train recruits to the Indian Forestry Service and it was done with the help of Germany, whose forestry traditions, like those of France, date back several centuries.

During the First World War, the Baltic once again became closed to Britain. Supplies from North America were also made exceedingly difficult by German U-boats and many a fine cargo ended up as driftwood on the shores of Connemara. So it became obvious at long last that a national forestry policy was desperately needed. At that time nine-tenths of Britain's timber supplies were imported.

It was now recognised that home-grown timber could no longer be left to the worthy efforts of individuals. But it was not until 1919 that the Forestry Commission was established by Act of Parliament, too late to help in the war-effort but at least able to plan for the future.

6 · Project Tree

The first tree planted by the Forestry Commission half a century ago, was at Eggesford in Devon. Now the Commission administers in Britain as a whole, nearly 3 million acres of land, of which over 2 million are under plantation.

In terms of a country as small as Britain this makes the Forestry Commission sound a large and important state enterprise. Every year Britain still imports £500 millions' worth of timber. In 1967 the Forestry Commission produced 1,200,000 tons of home-grown timber, worth rather more than £4 million. The Commission estimates that the production of softwood from Britain's forests should reach 4 million tons by 1980 and 5 million by 1985.

Though small, this is an essential contribution and represents a radical change. The harvest of home-grown timber will increase progressively as the plantations established during the past fifty years come to maturity.

Even so, it is necessary to set things in proportion. Only 7% of England and 9% of Scotland is covered by forest or woodland. These are among the lowest figures in Europe, only those of countries such as Denmark, Holland and Spain being comparable. Because of the smallness and dense population of Britain, it is doubtful whether the Forestry Commission can ever grow more than five million acres of forest. Even that would supply only one-third of Britain's requirements, which is the long-term aim.

In contrast, much of Europe is far more heavily timbered, both in proportion and in area. Twenty-nine per cent of Federal Germany is covered by forest, representing more than seventeen and a half million acres. Twenty per cent of France consists of forest and the French currently plan to afforest a further 23,000 square miles (equivalent to nearly half England), all of which will enable them to dispense with timber imports by the end of the century.

In Switzerland there are two and a half million acres of forest-land, more than 25% of the country. This is due to an enlightened policy going back to the fifteenth century. As for Sweden, more than half of her 111 million acres is covered by forest, while in the Soviet Union itself there are an estimated 8,000 million cubic metres of standing timber. This comprises nearly one-third of the total timber reserves of the world and makes Russia the world's biggest potential producer.

In the basin of the Siberian Yenisei alone stands coniferous timber four times greater in volume than the entire timber reserves of the rest of Europe and greater than those of Canada and the United States.

Even so, Canada has an almost unbroken belt of forest between 600 and 1,300 miles wide and stretching from the Atlantic to the Pacific. The United States has 750 million acres of forest, not far short of the total area of a country the size of India. Australia has 63 million acres of timber of all kinds and New Zealand expects to supply all her own requirements for wood and wood products by 1975.

However tiny British State Forests may be in comparison there is at least one of these in every rural county in Britain. Forestry Commission forests or plantations comprise nearly 380 separate units and range in size from the 50,000 acres of Kielder Forest in Northumberland to the 2,300 acres of Delamere Forest in Cheshire.

In addition to the Forestry Commission's own activities, an increasing amount of planting, some 40,000 acres a year, is being carried out by private landowners and farmers. This is done under the auspices of the Forestry Commission's 'Dedication Scheme' Not only is free technical advice and help given but also government grants to encourage tree-planting.

The forestlands under direct Forestry Commission administration vary greatly in character as in size. Some, such as Sherwood Forest, are old and new grafted together, as it were. Many huge oaktrees, which were probably acorns in Tudor days, still stand, even though the tree in which Robin Hood allegedly kept his larder is perhaps more legend than fact. But alongside these ancient woodland giants, grow 15,000 acres of conifers planted by the Commission.

Another even more historic forest which has been entrusted to the Commission's care is the New Forest, parts of which in spite of its name existed in primeval times. It contributes to the country's economy as well as to its traditions and even during the Second World War it provided twelve and a half million cubic feet of timber.

Though where possible the Forestry Commission has preserved

and extended the ancient forests, most of its woodlands are newly created. Halwill's 6,000 acres in Devon and the 37,000 acres of timber at Allerston in Yorkshire are both examples of moorland which previously never produced anything more than heather or bog-myrtle.

Land in Britain becomes increasingly precious as more is eaten up by new motorways, reservoirs, airports and towns. This means that what is available to the Forestry Commission consists, broadly speaking, of poorer land. This means in turn that it can only plant a limited number of 'traditional' English broadleaved trees such as oak, ash, beech and sycamore, which need a more fertile soil.

But apart from the fact that most of its activities have to be confined to hilly or moorland areas, the Forestry Commission has had to concentrate mainly on coniferous trees for another reason. The broadleaved trees or hardwoods take many years to reach maturity. As we have seen, the oak can be nearly two hundred years a-growing.

On the other hand, the conifers or softwoods are far sooner ready for use and, home-grown or imported, supply nine-tenths of all the wood Britain uses. Moreover, conifers yield a greater quantity of timber per acre in the hills of Britain than they do in the Northern European forests of Sweden, Norway and Finland.

So, as the Forestry Commission has to plan its output like any agricultural crop, it cultivates in the main those trees which will soonest give a return, for it has centuries of neglect to catch up on. The Forestry Commission's current afforestation programme embraces 450,000 acres scheduled to be planted by 1974.

7 · Planting by Spade and Helicopter

Any programme of afforestation entails a large amount of preparatory work, involving many different processes.

Massive land clearance may be planned, necessitating the use of crawler-tractors, mechanical diggers and ploughs far bigger than would be needed for ordinary farm-work. The land has to be ploughed partially, in strips five or six feet apart. This is primarily to prepare the soil, but also to discourage weed growth that might impede the young trees or increase fire-danger.

On some land, too, this ploughing is needed to prevent 'hard pan', a solid layer of soil that forms beneath the surface and stops the weak shoots becoming established.

In conjunction with the work of clearance, an elaborate deep drainage system will often have to be channelled. Even in an established plantation, or one that has been felled, this network of drain trenches is constantly being renovated. Trees can be 'drowned' just as they can be parched.

While the land is being prepared, a constant supply of seeds and seedling trees has to be maintained. The Forestry Commission has 2,000 acres of nurseries in Britain and the new growth is raised from seed obtained from specially selected trees, as a farmer will breed from the best stock available. Much of this seed is home-grown, but some is imported from countries as far apart as Corsica and Japan. The planning of seed-supplies is made more important by the fact that some trees, the oak, for example, produce seed perhaps once every five years.

Thirty million trees are needed every year for replacements in existing forests as well as for entirely new plantations. In fact, at any one time no fewer than seventy million of them are available for instant use. As the seedlings take from one to five years to become fit for planting out in the forest, they are obviously in very different stages of growth. The nurseries may therefore contain as many as two hundred million in the process of being raised.

3 *Bed of six-week-old Scots pine seedlings*

The first transplanting from the carefully tended, hand-weeded seed-bed is carried out when the seedlings are perhaps a year old and a few inches high. This is done by means of a 'lining-out board'. The seedlings are laid side by side on this, spaced out regularly by means of pegs and with their roots outside the edge of the board. When the board is loaded, a second board is laid over the seedlings and clipped to the first board.

4 *Filling a lining-out board with seedlings*

The load of seedlings can now be carried to the new planting ground, where the lining-out board is held vertically, with the little trees thus upright, while the roots are bedded into the soil. Then the clips on the boards are unfastened and the boards lifted off. Even this simple piece of automation has saved endless working in transplanting and also ensures better spacing and alignment.

In some cases a second nursery transplanting takes place later to strengthen the seedlings by giving them more root-space. Strong roots are essential, not only to help them draw sustenance from the often niggardly soil, but also to help them, as far as

5 Seedlings may be 'undercut' so that the trimmed roots grow more strongly and a second transplant is avoided

possible, to withstand storm-damage. But not even the sturdiest trees can withstand some weather, however well rooted they may be. In a few hours one day in 1968, 8,000 acres of forest property in Scotland were destroyed by a hurricane that carved a fifty-mile swath.

While the seedlings are in the nursery stage the most rigorous weeding has to be carried out. In order to avoid disturbance by hand-methods, this is often done by spraying carefully selected weedkillers. Spraying is also used against possible diseases such as blight and canker.

The main transplanting of the young trees into the forest-land takes place when they are between three and five years old. Frost-free periods in autumn and winter are the usual time for this, when the trees are dormant.

One method is called turf planting, in which the soil is turned up by plough and the trees planted, by means of a spade, in the upturned sod. But the most usual and certainly the easiest method, as the ground does not require any preparation (other of course than the original clearance ploughing which we mentioned), is 'notch' planting.

In this, two notches forming a right-angle are cut with a spade, the ground is prised up with the second cut and the young tree is thrust into the soil while the top sod is then pressed down firmly with the worker's foot. The trees are planted between four and six feet apart, with an average density of 1,750 to the acre, and one man can plant as many as a thousand trees a day by the notch method.

In Britain, hand-planting is the sole method. But in a country such as the Soviet Union, where tens of thousands of square miles are involved, successful experiments have been carried out in the automatic planting of trees.

6 *A mist-blower is used to apply spray to an area of woody weeds*

7 *Notch planting*

8 Seeds for timber crops are sown by helicopter in the State of Washington, USA

This is done by means of a system which feeds seedlings into the planting arms of a 'forest-laying' machine. The seedling-holders are loaded on to a machine worked by a single operator who can load thirty holders with six hundred seedlings in less than twenty minutes.

From the holders the seedlings are fed in a continuous flow into the planting arms which handle each small plant with exact care and place it in its appointed situation. The Russians hope that when this system is fully developed, one operator will be able to plant twenty acres of seedlings in a working-day.

In contrast the Americans have in many cases taken to direct tree-seeding which is carried out by helicopter, the first experiments having taken place as far back as 1940.

In this method, which is also used in New Zealand, and to a limited extent in Australia and Tasmania, marker-flags guide the forester-pilot to the area in question. While he flies at two or three hundred feet and between fifty and sixty miles per hour, seed is sown from two side-hoppers. An even distribution of one pound an acre over a ninety-foot path is ensured by means of a special meter which controls the power-driven 'slingers'.

In the case of some species of tree there are as many as 40,000 seeds to the pound, which means that the pilot can sow something like 25 million seeds to the square mile. Needless to say, nothing like this number eventually grow into trees.

One major precaution that has to be taken in this aerial sowing is to protect the scattered seeds against rodents and birds. The birds are repelled by coating the seeds with aluminium paint which makes them suspicious. The rodents, mainly mice and voles, are put off by the use of Endrin, a chemical which gives the seeds an unpleasant taste.

8 · Guarding the Crop

The forester cannot sit back once the trees have been planted. He has to wait a good deal longer than a farmer for his harvest, but, like the farmer, he has to tend the crop continually.

Inevitably, the following year after planting-out it is found that some seedling trees have perished. An average of 10% have to be replaced. Not all the surviving trees will be allowed to reach anything like full maturity. A certain amount of thinning is always carried out in the new forests to allow the bulk of the trees proper space in which to grow.

Occasionally, thinning is done by deliberately poisoning the trees. This is carried out by means of a power-driven auger, a New Zealand invention, which is fed with poison carried in a plastic container above the portable motor.

But the thinnings are not all wasted. Young spruces can be sold as Christmas trees, while in a few years' time other conifers are sturdy enough to be used for pit-props, fencing-poles, and similar purposes. Although the big final crop is what really matters, a plantation will be producing timber for many years beforehand.

It is occasionally necessary to thin out trees for another reason. Frost can wreak great havoc among young trees, and in extreme conditions the danger can sometimes be lessened by thinning out the plantation. This enables the air to move more freely, which reduces the chance of frost-damage, for frost is worst in a still atmosphere.

Until the trees are well established, a constant and vital chore is weeding. Honeysuckle and ivy may strangle or distort trees, but the main 'noxious herbs' are bracken and brambles – and grass of various kinds. The weeding is carried out chiefly with the sickle or cutting hook and at first has to be done with great care. The young trees are often lost in the surrounding weeds and can easily be lopped off by mistake.

Experiments are taking place on some Forestry Commission land which will, if successful, lessen this task of weeding. Instead of completely clearing the 'slash', the ruck and rubble of trees that have been felled, these are being left and the new young trees are planted in between. The debris not only protects the young trees but also helps to suppress the weeds.

But at least where weeding has to be done it is made simpler by planned, uniform planting. This of course is also designed to facilitate other operations such as thinning and eventual harvesting.

A careful check has to be kept on fungi of different kinds. Some fungi actually help in stimulating the growth of a tree; others are a hindrance. Some species of tree suffer a specific fungus which attacks them alone. The 'elm-disease', and the 'sooty bark' disease which attacks the sycamore, are examples of this. So, too,

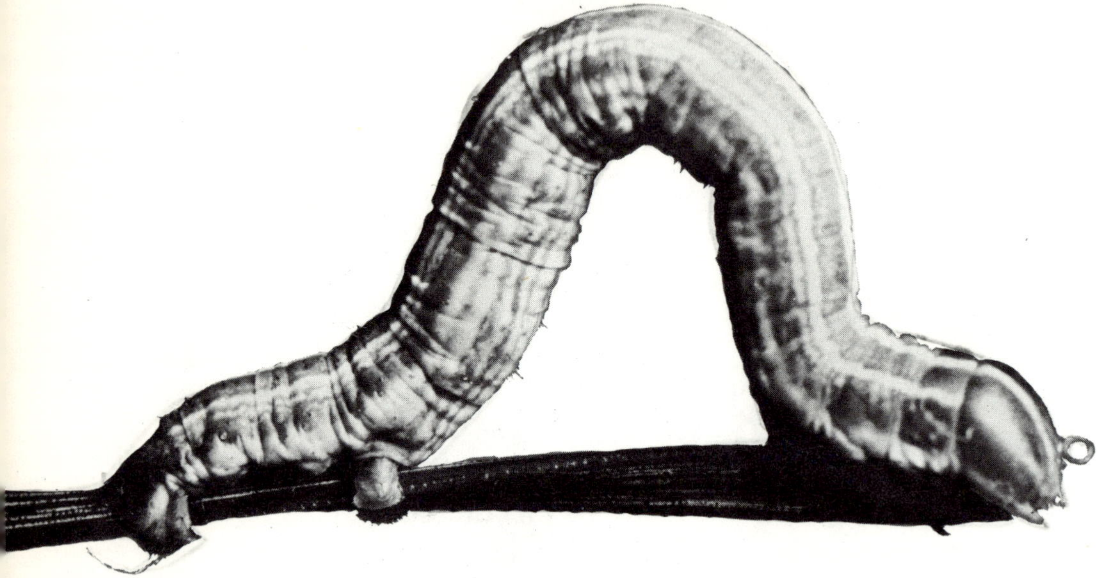

9 Larva of the Pine Looper moth which attacked Cannock Chase, Staffordshire

is 'oak wilt', another fungus disease which caused immense damage to oak-trees in Arkansas and Ohio, while 'chestnut blight' killed off most of the American chestnuts.

Insects, too, can make the forester anxious. One species will seriously harm trees such as the Douglas fir, whose needles become covered by a white woolly substance in which the insects live while they are sucking away at the life-juices of the tree.

The spruce sawfly damaged huge areas of Canada's forests.

10 Ponderosa pines in California, destroyed by tree-killing insects

This was only combated by the artificial introduction of a suitable parasite, which had to be imported from Europe, where the sawfly had originated.

One of the most difficult insects to deal with is the bark-beetle, which lives under cover of a tree's bark and can cause widespread damage. Sometimes the only way of wiping it out is by means of what American foresters call 'sanitation salvage logging', which is the wholesale removal of infected trees and their neighbours.

Although their use is debatable, chemical insecticides or fungicides can be sprayed by farmer or fruit-grower to combat various pests. But because of the large areas involved and the size of his crop, this is usually impracticable for the forester. Apart from the hand-spraying of the smaller, more compact nurseries of tiny seedlings, only in one case has the Forestry Commission resorted to spraying a forest.

This was in a year when the larvae of the Pine Looper moth suddenly irrupted to such an extent that they were rapidly defoliating part of Cannock Chase. The problem became so serious that DDT had to be sprayed from aircraft. Even in America, where chemicals are used so widely in agriculture, they are infrequently employed in forestry.

Animals in a forest, being generally more evident and less numerous, can be more easily controlled. But here it must be said that the modern forester pursues as enlightened a policy as possible towards nature. It is realized that too often in the past the balance of nature has been upset by indiscriminate interference. To take but one example: the artificial introduction of the rabbit into Australia during the last century caused immense problems because there were not enough natural predators to keep its numbers in check.

A few birds such as greenfinches and chaffinches will attack seedling conifers, while in Scotland the giant capercailzie and the black grouse feed to some extent on pine-shoots. But many birds act as a natural deterrent to harmful insects, while wood-pigeons and jays can even help in the regeneration of a deciduous forest by spreading acorns and beechmast, through their droppings.

It is mammals that often cause the forester the greatest concern. In America, bears for example, will rip open the bark of trees to get at the sweet cambium inside.

11 DDT fog is sprayed over the forests of eastern Quebec in an effort to salvage a million acres of woodland

One of the first things that must be done when a new Forestry Commission plantation is created, is the erection of rabbit-proof wire all round the perimeter. Since the deliberate introduction of the disease myxomatosis into the country, the number of rabbits has dropped enormously. Even so, there has been a partial revival of these destructive creatures and they are still the arch-enemy of the forester in Britain as in Australia.

Special high fencing has to be erected in forestland near which deer exist. They can do damage to trees not only by browsing but also by fraying – that is to say rubbing their antlers to get rid of the 'velvet' which covers the new year's growth. The deer population of Britain has increased in recent years and comprises several different kinds: the majestic red deer, the fallow, the roe and the artificially introduced sika or Japanese deer.

But the Forestry Commission is by no means automatically hostile towards deer. In some forests, far from being kept out, herds of deer are allowed to exist freely. The numbers and quality are controlled by experts, who keep watch at certain times from specially constructed look-out towers.

This control is exercised not only to keep the herds at the best level of quality by 'culling' inferior animals, but also because the foresters know they have a responsibility to farmers outside the forests in question.

Two mammals that cannot be fenced against are the grey squirrel and the tiny field-vole. Both of these cause a great deal of damage to young trees. The grey squirrel (which was first introduced into Britain from America nearly a century ago) was formerly very prevalent and largely replaced the native red squirrel. In the past twenty years its numbers have been greatly reduced by shooting, encouraged by a two-shilling-a-tail bounty paid by the Ministry of Agriculture. The grey squirrel is another example of interference with the balance of nature. Its prolific breeding habits, together with the fact that so many natural predators such as hawk and owl had been killed off in the interests of game-preservation, led to its getting out of control.

As for the field-vole, it is difficult to do anything – except to encourage its natural enemies. Too often in the past a particular species regarded as a pest has been killed off, only for Man to

An animal 'pest'

12 The grey squirrel

13 A tree trunk showing squirrel damage

14 Badger gate in a Forestry Commission wire fence

find that the creature it preyed upon immediately became an even worse pest.

So the forester is glad to see birds such as hawk and owl and mammals such as fox and badger leading their normal lives. Indeed, as far as the badger is concerned, the forester regularly builds special little flap-gates in his fences so that this ancient British animal can come and go at will, without having to dig under the wire.

9 · The Smoke-jumpers

Disease and insects do more total, gradual damage, but the most terrible hazard of all in forestry is fire. Scrub-covered land, however, may be burned off deliberately, under strict control, preparatory to new planting, as in New Zealand, for example.

Fire is most dangerous of all in coniferous forests. Broadleaved trees do not burn so easily, nor do they grow so closely together. But conifers, as well as growing or being planted more densely, contain a great deal of resin which adds to their inflammability. Yet in a country such as Sweden, whose forests consist almost entirely of conifers, a total of only 0.1% of forestland has been damaged by fire in the last ten years.

In Britain, where the climate is fairly humid, the danger of fire is also less prevalent. Even so, in some years forest fires in Britain have caused as much as £50,000 worth of damage, and entail an expenditure of a quarter of a million pounds each year on protective measures. Forest fires can be started by natural phenomena such as lightning, but human beings are undoubtedly the chief culprits, through carelessness about camp-fires or cigarette ends.

So when a new forest is planned, fire-risk is always taken into account. Blocks or compartments of trees are usually confined to about twenty-five acres each. Fire-breaks, wide glades where no trees at all are planted, must be left at intervals within a plantation. In some cases these fire-breaks are even ploughed to keep down grass which, when dry, acts readily in spreading flames.

Round the edges of the block, wider fire-breaks are left, so that as far as possible it is isolated from neighbouring plantations. In addition, tall watch-towers are built in every forest area and in dangerous periods these are manned day and night. The guards can pin-point the first tell-tale curl of smoke by means of field-glasses and compass and raise the alarm by telephone.

Provided an outbreak is reported in time, it can be dealt with by a line of men beating with the familiar firebrooms of birch-brush or rubber flaps, which may be seen in racks on every forest road. If it gets worse, the forestry headquarters' own mobile pumps, augmented by local fire-brigades, have to go into action, where roads and water-supplies allow.

But if a fire gets out of control, particularly in an adverse wind, then it may be necessary to fell trees in its path or use mechanical diggers to dig trenches. In drastic cases it is even necessary to start another fire to stop the main one. This is known as counter-firing, when a fire is purposely started ahead of the real fire which thus runs out of fuel when it reaches the deliberately burned area. This is a standard practice in Australia, for example, where incendiary material may be dropped from the air.

In countries with a dry climate, fire can destroy the work of centuries, in a matter of hours or days. Horrifying fires have occurred in recent years among the pine-woods of the Riviera and the Landes in France. In Australia and Tasmania, bush-fires are an almost annual disaster – and the 'bush' often includes valuable timber. Thousands of square miles have sometimes been devastated when in time of drought, fires have swept across the land, driving people from their homes and destroying everything in the way.

In America, too, some of the most devastating and spectacular forest fires have occurred. Periods of drought, when 'fire-weather' prevails and when the forests become virtual tinder-boxes, are a time of apprehension. All logging operations are stopped. All visitors – hunter, hiker, fisherman, camper – are forbidden to enter the forests. The results of forest holocausts are too well known to the Americans for them to run any risks.

Calamitous fires still occur, as in Alaska in 1966, when nearly three-quarters of a million acres (equivalent to the East Riding of Yorkshire) were destroyed. However, thanks to modern techniques and equipment, the annual loss through forest fires in America has been reduced by nearly 90% during the last quarter of a century.

Not so long ago, forest firefighters in America and Canada had to rely almost solely on pack-horses to get their equipment to a threatened area. Nowadays, because of the extensive building of forest roads, trucks equipped with radio-telephones can maintain regular patrols; bulldozers and tractors loaded with fire-fighting material, including giant tanks of water, can be sent into the danger zone when a fire breaks out.

In addition, aeroplanes and helicopters are being used increasingly. Not only can air-patrols help in spotting fires, but converted bombers have been used to spray an area with special chemicals to retard a fire.

15 *The local fire brigade is called in to control a plantation fire*

Another important development is the technique of 'fire-jumping' or 'smoke-jumping'. In this, squads of trained men, clad in fireproof clothing, are parachuted, together with pumps and water-supplies and other essential gear, on to a suitable location ahead of a fire. The speed with which these smoke-jumpers go into action can make a vital difference between a localized outbreak and a major disaster.

Forest fires are a disaster in every way. Valuable timber is destroyed, together with the wildlife that inhabits it. Irreparable damage can be done to the essential, nitrogen-bearing humus of the forest floor, organic material, which is constantly being renewed by leaves and twigs and constantly decomposing.

But a forest fire can bring human tragedy as well. The Americans have good reason to be apprehensive of any forest fire. One of the worst of such catastrophes occurred in 1871, when the entire lumber town of Peshtigo in Wisconsin was wiped out. Nearly 1,700 men, women and children were roasted alive. Nearly four million acres of forest were laid waste.

The cause of that disaster: some railway workers were burning off a pile of brushwood. Their bonfire got out of control.

10 · Forests before, Deserts after

The value of the tree as timber is obvious: that is to say wood as a material, apart from the various ways in which it can be processed. But even if a suitable substitute for timber had been discovered, trees and forests would still be vital to the welfare of mankind. It is by the very process of growing that the tree individually, and the forests collectively, perform a most important function.

During its growth, a large tree takes in by means of its roots many tons of water. It has been calculated that to make a hundred pounds of cellulose, which is the main constituent of wood, a tree needs more than half that amount of water.

Yet the tree retains only a tiny fraction of this water, perhaps 1%, in the growth process of the cells. The remainder is given off through its leaves into the surrounding atmosphere.

These quantities are significant. Even a small tree, such as a fruit-tree, discharges four gallons of water into the air every hour. An oak tree, with perhaps a quarter of a million leaves, will discharge more than a thousand gallons a day in summer-time.

When such amounts are translated into terms of a forest, even though the individual amounts vary according to the species of tree concerned, the total result must obviously be vast.

The tree can be seen as a kind of pump with a dual purpose. It is constantly sucking up water from the ground, which would otherwise become far more loosely composed. At the same time it increases the humidity of the surrounding land, which would otherwise be much more arid.

Furthermore, trees and forests act as a breakwater against the full direct force of rainstorms which, especially on hillslopes, would otherwise wash away much of the topsoil, both in the immediate vicinity and lower down. Not only this, but the rich humus comprising the floor of the forests acts as a sponge. It absorbs far more water than the bare earth would be capable of coping with.

What is not required by the trees themselves is filtered through to the subsoil in an orderly fashion, feeding the springs and ultimately the streams and rivers of the region. Apart from protecting the soil from the uncontrolled effect of rain, the trees protect it also from wind-erosion, for wind can blow away naked or overworked soil as surely as water can wash it away.

The considerable influence trees and forests have is plainly evident from this. They affect the soil in which they grow and the surrounding land, the atmosphere and also the whole climate of the immediate region.

Trees and forests are an integral part of the delicate balance of nature. Many kinds of natural phenomena, such as earthquakes, volcanic eruptions, prolonged drought, the movement of glaciers, and so on, have all at times affected or upset this balance.

Nothing can be done about such happenings. But some of the

worst instances of interference in the balance of nature could have been avoided. For they have been caused by Man himself, as often as not against his own ultimate interests.

Man, as well as being the most intelligent of the world's inhabitants, is also the most destructive. From the proverbial but once real dodo to the teeming passenger pigeon, from the magnificent blue whale to the graceful oryx, history is strewn with examples of animals man has either wiped out or brought to the verge of extinction.

Far more disastrous, simply from his own material point of view, has been his treatment of the tree. The French writer, Chateaubriand, summed the matter up: 'Forests precede civilization, but deserts follow it'. This, unfortunately, has been borne out many times all over the world.

Parts of the Sahara were once green and fertile. Lime trees, holm-oaks, alders, cypresses, flourished there. In what was Palestine, much of the land consists of barren rock where once the almost fabled cedars of Lebanon flourished. Until recently perhaps four hundred remained, although today a scheme is underway to save them from extinction.

In India waterless ravines have replaced the once famous sal forest, that grew thickly along the river Jumna. A Chinese river such as the Hoang-Ho has been known to sweep away 500 million tons of fertile soil annually, because of the destruction of the country's forests. In America disastrous dust bowls such as the Tennessee Valley were created by the indiscriminate felling of trees. The wholesale cutting of trees in parts of West Africa resulted in a serious change of climate.

The list is endless. But it was not only a question of the forests being plundered for timber. Vast tracts of land have been ruined by wanton grazing, particularly on the part of one animal: the familiar, domesticated goat. Many specialists in the field of natural resources would go so far as to condemn the goat as Man's worst enemy. It was largely due to the goat that much of the land bordering the Mediterranean was denuded of trees.

So, through indiscriminate felling and grazing, once profitable land was left at the mercy of rain and wind. The wind not only blew away the unprotected soil but also helped the sand of existing deserts to engulf yet more territory.

In addition, by allowing erosion to occur, Man also increased the likelihood of another danger. The masses of soil washed away often silted up neighbouring rivers to a catastrophic extent, sometimes far away from the original forests. As a result, the riverbeds were raised considerably and became liable to flood. This is yet another disastrous result of the wholesale destruction of the trees.

Fortunately, nowadays, because of our awareness of the urgent need for conservation, something is being done to redeem the mistakes of the past.

16 Shelter belts are planted in southern Texas to protect farmland against wind erosion. It is usual to have ten rows of trees per belt, with the tallest in the centre and shrubs on the edges

17 Transplanting and watering eucalyptus and pine seedlings at a tree nursery in Kais, Algeria, as part of a soil conservation and reclamation project

The Americans have realized the value of trees not only for timber but also for soil protection. Likewise the Russians are countering wind-erosion in the Kara-Kum desert of Turkmenia by planting large areas with the drought-resistant cherkez tree. In Kazakhstan, too, they plan to plant broad shelter belts of trees to influence the climate and so protect some three million acres of farmland. Spain is beginning to rehabilitate her parched 'Meseta' by extensive afforestation.

North African countries, helped and advised by the Food and Agriculture Organization of the United Nations, are planting eucalyptuses and poplars along the fringe of the Sahara to push back the desert. China is in the process of building a new 'Great Wall'. But this is a Green Wall, consisting of a six thousand mile belt of forestland to form a bulwark against the wind-eroded wasteland and to improve climatic conditions.

An extremely important modern development in which the tree is being used to rehabilitate desert or barren land is known as 'three-dimensional forestry'. It has received considerable attention through its application during the last decade in the Limpopo Valley in Africa, but was originally developed in Japan.

In certain mountainous regions the Japanese were faced with the problem of soil erosion because of wholesale forest destruction in the past. But instead of simple reafforestation with timber trees, they decided to use nut trees. These not only helped to stop erosion and provided timber, but also supplied fodder for animals, which in turn provides food for human beings.

In addition, the forest plantations in question were divided by carefully regulated grazing strips. This meant that supplementary feeding-stuffs from the trees were ready to hand, while the natural manuring of the pasture land led to its rapid improvement.

Israel provides another striking example of how modern man is attempting to make the desert bloom again with the aid of trees. In the Negev, in the barren south, the rainy season sometimes lasts no more than a week and produces an inch of rain. Six inches of rain constitute a wet year, obviously insufficient for the cultivation of trees.

On the other hand, dew falls plentifully throughout two-thirds of the year. So Israeli scientists have successfully experimented in collecting the dew on polythene sheets spread round the sapling trees. The resultant moisture – ten pints a month from each small sheet – is fed along a miniature gutter to the roots of the tree in question, tamarisk, pine, cypress and eucalyptus.

For a final example of how Man has called in the tree to help him, in Switzerland belts of trees are being planted especially to minimize the danger of avalanches in vulnerable areas.

11 · The Life of a Tree

So far the discussion has been of trees and forests in general. But of course it is the individual tree that matters and it is one of the first jobs of the forester to understand the system by which a tree lives and grows.

To begin with, the seeds of different trees are highly intriguing. It might be thought that the bigger the seed, the bigger the tree. But this is not so.

A striking example of this discrepancy between seed and tree is the magnificent redwood of California. This tree, which attains a height of three hundred feet and a weight of nearly one hundred tons, originates from a seed less than three-quarters of an inch in length. Three thousand redwood seeds would tip the scales at – a single ounce.

To begin with, the warmth and moisture of spring trigger off the chemical processes inside the seed as it lies dormant in the soil. Then a tiny thread of a root thrusts downwards in search of support and sustenance. Correspondingly, the 'tree' itself begins to grow upwards in the form of an equally fragile stem.

At first, the woody content of a seedling tree is nil. The seedling consists only of its vascular system (what might be called its blood-stream) circulating round the outer part of the stem. But wood starts to develop continually with the growth of the tree and is formed by what is known as cambium, a ring of cells completely encircling the stem and always forcing themselves outwards.

This cambium layer is of fundamental importance to the life of the tree, for it is the formative tissue and develops uninterruptedly in two sets of cells. Those on the inside are the wood cells and become first sapwood, carrying sap *upwards*, and later heartwood, the two making up the timber-content of the tree. Those on the outside are known as phloem or bast, and form the *downward* food-carrying system.

While the cambium or wood cells of trunk and branches expand continually, producing increasing quantities of solid heartwood, this same outward action breaks open the phloem. It is also replaced as the outer layer by the 'cork' cambium which becomes the bark or protective covering of the tree. This really is cork, and it is from the bark of a species of oak growing in Spain and France that the cork we use for such purposes as bottle-stoppers and life-jackets is obtained.

On certain trees it can plainly be seen how the bursting effect of the outward-pushing cambium takes place. Because of this action the bark is being constantly renewed, though occasionally a tree can become 'bark-bound', in other words its growth is impeded by the tightness of its bark.

The growth process of cambium and phloem depends on the sustenance the tree is able to absorb in different ways. This is made possible by a two-way traffic.

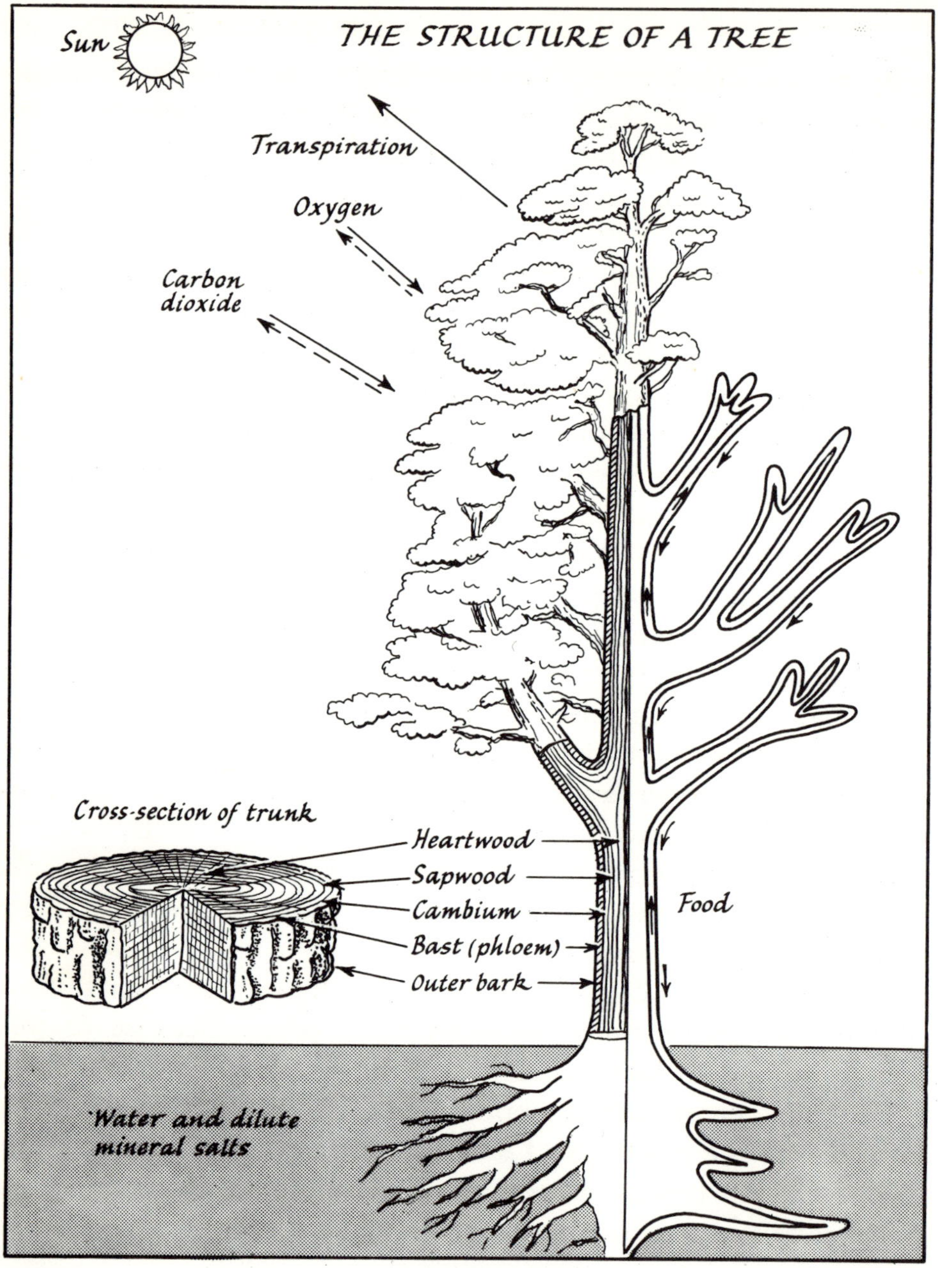
THE STRUCTURE OF A TREE
Sun
Transpiration
Oxygen
Carbon dioxide
Cross-section of trunk
Heartwood
Sapwood
Cambium
Bast (phloem)
Outer bark
Food
Water and dilute mineral salts

Fig. 5

By means of thousands of tiny hairs or fibres, the roots draw up water containing various chemical ingredients acquired from the soil. An essential element in this process is the countless microscopic bacteria, algae and fungi. These minute organisms are vital in speeding the decay of a forest's debris which would otherwise eventually choke it. In so doing they also release basic chemicals and minerals such as phosphorus, sulphur and nitrogen, which help the individual tree to grow.

These chemicals and minerals are contained in the water or sap which rises through the whole structure of the tree to its outermost shoots. As the tree grows, the original sapwood dries out and hardens, largely through the influence of tannin, a vegetable substance contained in the tree and which has very astringent qualities. It becomes true wood or xylem, the heartwood which is the real timber but which is in effect now dead.

In fact, the living parts of the tree consist only of the ends of the roots, the leaves and buds, and the sapwood and the phloem. The sapwood and the phloem perform the continuing function of carrying the life-giving sap to create the new sapwood the tree is pushing out all round its girth every year.

It is often imagined that sap rises and falls according to the seasons. But there is always sap in the tree; otherwise the tree would die. It is true however that in spring the sap does flow more strongly.

Whilst the sap flows upwards through the tree, the leaves perform a complementary function. Through microscopic pores they absorb carbon dioxide from the atmosphere. This merges with the sap through the aid of the sunlight and the chlorophyll, or green colouring matter, in the leaves themselves. This is a process called photo-synthesis.

The result is a supply of carbohydrates contained in a sugary solution which travels back down the tree, by way of the bast, and recharges the roots with the necessary energy to keep the system working.

An idea of how the food-distributing system of the tree works can be obtained by examining a tree-stump or a sawn log, preferably with magnifying glass or microscope. From the centre to the circumference are a series of 'medullary rays'. These are the horizontal food-bearing and food-storing vessels, for the tree, except for its dead centre, needs to be fed throughout its body and not merely up and down.

Also visible are the concentric rings spreading out from the centre which was the original tree. The wood becomes suddenly less brown at a point near the outside because of the chemical change which has taken effect at the heart of the tree.

Every year, and especially in spring during the time of fastest growth, the tree develops another layer of sapwood. This shows on a cut log in the form of an additional circle.

As the sap is most active in spring, the cells of a growth-ring formed then are much larger, so that part of a ring is far more pronounced than the part of a ring produced by the slower summer growth. In winter there is no growth at all. These annual rings amount to what might be called the birth certificate of the tree.

As well as being the clue to the age of the tree, these growth-rings are by way of being the chronicle of the tree's career. An expert botanist could tell by studying an old tree what kind of weather occurred during the year of Trafalgar or the coronation of Queen Anne.

But these growth rings are only reliable in temperate countries where there is a sharp difference in the seasons. They do not apply in the same way to tropical conditions.

Many records have been claimed about the longevity of trees. In Teneriffe there was the Dragon's Blood Tree (so called because of the colour of its resin) which was reputed to be six thousand years old. 'The Big Tree of Tule' was claimed by the Mexicans to be even older.

But probably the most authentic example is the Bristle-cone pine of the White Mountains of California. This, it has been scientifically established, is some four thousand years old – which means it was growing when the Ancient Egyptians were building the pyramids.

Research on tree-rings and means of ascertaining the age of timber is being sponsored in Britain by the Natural Environment Research Council. This work is based on the scanning of radiographs of wood cores by means of a microdensitometer and will be invaluable not only for tree-study, but also for archaeologists in determining the age of wooden objects.

As for the growth of trees themselves, a highly interesting study is being carried out at the East Malling Research Station in Kent, in this case with particular application to fruit-trees. Here the root growth is actually being watched by means of an underground root observation laboratory. This is equipped with plate-glass windows fixed against the soil and allows the use of different kinds of testing and measuring equipment such as binocular microscopes and soil-moisture tensiometers.

12 · Evergreen or Deciduous?

In the Forest Products Research Laboratory at Princes Risborough in Buckingham, there is a reference library stacked to the ceilings with shelf after shelf of items.

It is not a library of books, as might be thought at first sight, but a collection of specimens, 28,000 altogether, of all the different kinds of timber in the world. This is a mammoth total, though out of all the infinitely varied kinds of tree that exist, only three or four hundred are used commercially.

As far as timber is concerned, the two chief categories of tree are deciduous and coniferous. Most broad-leaved trees are deciduous, and they generally grow in the temperate parts of the world. They shed all their leaves at the same time during a definite season which is mainly autumn. In the tropics they shed their leaves gradually through the year, so that they are never completely bare.

The second category, conifers, are mostly evergreens. In

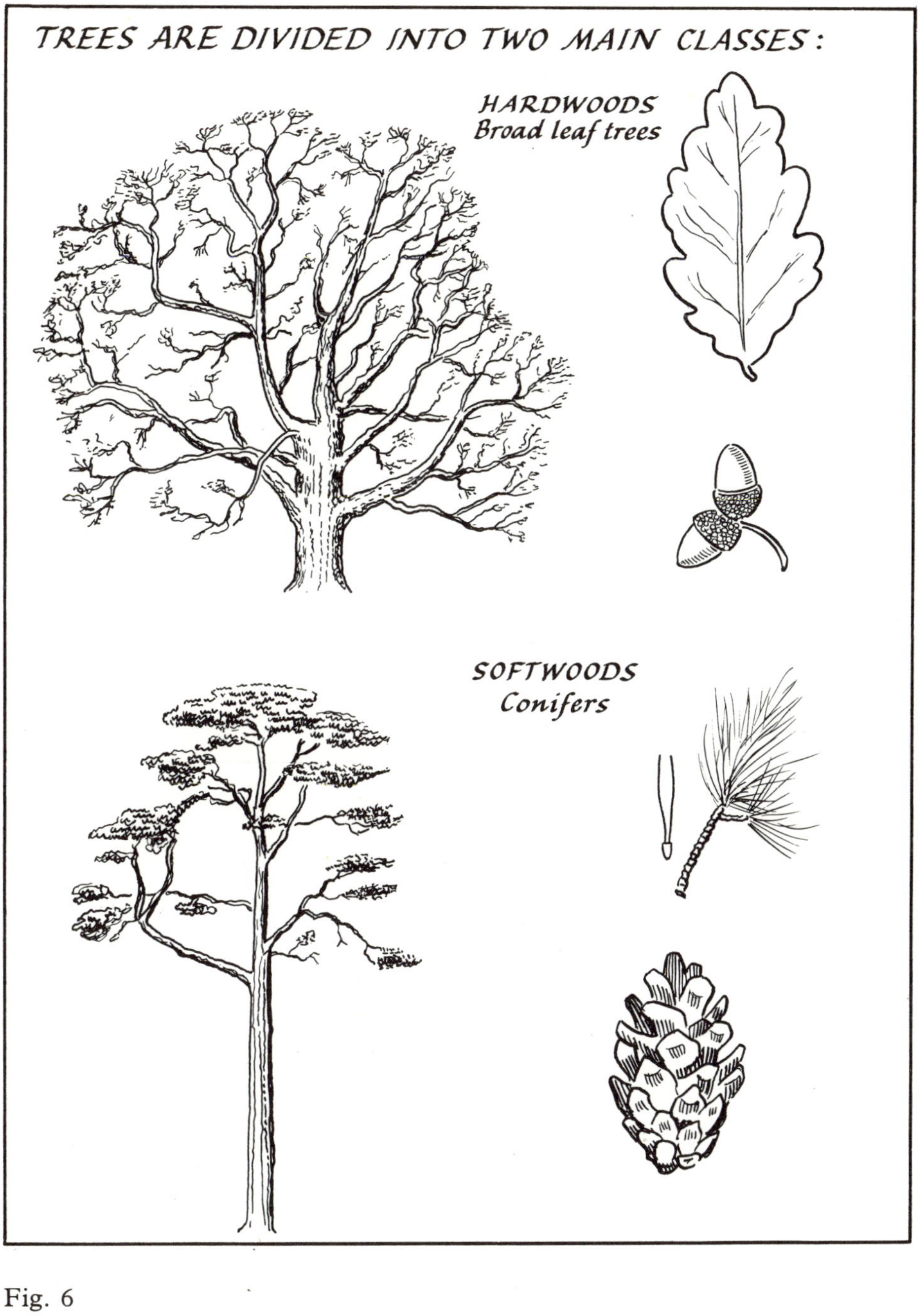

TREES ARE DIVIDED INTO TWO MAIN CLASSES:
HARDWOODS
Broad leaf trees
SOFTWOODS
Conifers

Fig. 6

principle these have needle-shaped leaves and they, too, shed them gradually throughout the year and are never bare.

These two categories correspond, again broadly speaking, with the main division between hardwoods and softwoods respectively. Timber from deciduous trees tends to be harder than that from conifers. Softwoods, as their name implies, are usually easier to work in carpentry and in other processes.

But there are contradictions in all this. For instance, one of the softwoods, the larch, is not an evergreen like its relations, the spruce or the pine. It is coniferous, bearing cones which develop from the beautiful 'larch-roses', the red female flowers. Yet it sheds its tufts of 'leaves' in autumn exactly like a deciduous tree. Again, some of the hardwoods are evergreens, such as the holm-oak and the holly, which shed their leaves through the year.

The tropical balsa tree and the yew are two other trees with opposing characteristics. The balsa, from which was constructed the famous Kon-Tiki raft, is a hardwood, yet as model aeroplane enthusiasts know, it is almost as soft and absorbent as cork. The yew, nominally a softwood because it is coniferous, is in fact one of the hardest of timbers.

However, it can safely be assumed that in general hardwoods are broad-leaved and deciduous and their timber is harder in texture, while softwoods are coniferous and evergreen and their timber is softer and more workable.

Hardwoods are a great deal more varied than the softwoods. This is exemplified by the fact that in the case of the oak alone, there are more than three hundred kinds throughout the world. But though the softwoods are far less varied in kind, commercially the forests they comprise cover a far greater area than do the hardwood forests. Softwood products of all kinds account for the greater part of the world's timber products.

As for the forests of the world, they vary greatly, too, according to climate and latitude and also altitude. The world is divided roughly into zones in each of which forests of the same kind tend to concentrate right round the globe. But in some regions hardwoods and softwoods do mix.

Far enough north or south, no trees exist. Inwards from the polar regions, in other words towards the equator, the trees begin gradually to gain ground. This applies more to the northern hemisphere, where there is more land within the arctic circle than in the antarctic.

In the north of Lapland, for example, the tiny, gnarled, alpine birch struggles for existence. At first it almost crawls along the ground, then later it starts to stand upright.

But real trees and real forests are still a long way off. Many of the northern Lapps have to travel considerable distances to buy timber which they ferry home downriver in the form of rafts for building purposes.

Such timber comes from the 'northern temperate' forests of the coniferous or softwood zone. The edge of this overlaps the arctic circle and stretches south as far as parts of the Soviet Union and the United States and includes much of Britain as well as all Scandinavia and the Baltic littoral.

There is a southern equivalent to this, but of far less significance for the simple reason that the only land consists of islands such as Kerguelen and Macquarie.

The next two zones match each other more closely. The northern belt embraces most of Britain, Europe including the Soviet Union, parts of Asia, including Japan, and central North America, including Mexico. Its southern counterpart includes South America below the Amazon basin, much of South Africa, Tasmania and New Zealand.

These two temperate hardwood zones consist of forests made up either of hardwoods only or mixed hardwoods and softwoods.

Last of all, situated between the tropics of Cancer and Capricorn, in other words a huge belt through which the equator passes, is the zone of tropical forests. This takes in Southern Asia, Indonesia, Australia, central and northern South America and most of Africa.

Comparable with the way in which trees and forests vary according to latitude – and therefore climate and soil – is the way they are affected by altitude. In the case of a mountain in the northern zone, the trees would continue to be coniferous, but they would eventually disappear altogether above the 'tree-line', that is the maximum height at which trees could grow in the latitude in question.

But in the case of a mountain in the tropics, altitude would provide a microcosm of all the different latitude-zones in the world. Near the base, tropical forest would exist. Higher up, temperate hardwoods would take their place. Finally, because altitudinal conditions would resemble the latitudinal, conifers only would grow.

These in turn would, as in the far north, become stunted and gradually peter out and the mountain peak would emerge into a world of snow and cloud.

13 · Hardwoods

Although far more conifers are nowadays being planted in Britain, it is the hardwoods or broad-leaved trees that are most typical of the country and the Forestry Commission has planted as many as thirteen million of them in a single year.

The majority of these hardwoods are native to Britain or, in the case of those which have been introduced, of several centuries' naturalization. First and foremost is the oak, although in fact, much oak timber used commercially in Britain comes from countries such as Poland and Japan. Some 800,000 acres of native oakwoods remain and the Forestry Commission has added to these by planting six million oak seedlings.

The marvellous strength and durability of the oak are such that it is still used extensively in heavy industry. In addition it is still greatly valued in the making of items which need to be completely reliable, such as heavy-duty railway wagons, barrel staves and ladders.

The ash is a tree the strength of whose timber approximates to that of the oak and of the American hickory, though in a different way. It is supremely resistant to shock in every form, being both tough and resilient. That is why it is the most suitable wood for a variety of purposes, from vehicle bodies to hammer shafts. It is commonly used, too, for the handles of gardening implements such as spade and fork.

The ash could also be called a very sporting wood. It is used for a great deal of games equipment, such as tennis rackets, hockey sticks, cricket stumps, skis, oars and gymnasium gear. But of course the tree most closely associated with one of England's national games is the willow, which is almost synonymous with the cricket bat.

There are nearly one hundred kinds of willow growing in Britain and they are used for a multitude of purposes, including matches, knifeboards, honeycomb frames, artificial limbs, balls for polo or coconut-shies, weatherboards for barns and the bottom of carts and lorries.

But out of all these different kinds of willow, only one is suitable for making the cricket bat and it is appropriately known as the cricket-bat willow. Like the ash, it is able to resist considerable shock without splintering. Yet at the same time it must be light in weight: a cricket bat weighs only a few pounds.

The beech, whose majestic grey trunk and glorious crown make it one of the handsomest of trees, was long ago more valued for its rich mast for fattening pigs than for its timber. Today it is Britain's biggest hardwood 'crop' after the oak, and nearly 200,000 acres of beechwood remain, including six million trees planted by the Forestry Commission.

Because it is strong yet easily worked, the timber of the beech is put to a number of minor uses such as toys, shoe-heels, rolling-

18 Hardwoods are used in the manufacture of the traditional Windsor armchai
Here one is being prepared for export to the United Sta

pins, platters, ice-cream spoons and clothes pegs. The Danes use it for another reason: more than most kinds of wood it is free of taste and smell, which makes it ideal for the million butter casks they need every year. But beech has mainly been associated with furniture-making, partly because of its smooth grain and partly because it is better suited for indoor uses and does not stand up well to exposure.

We use the word 'bodge' to indicate something clumsily done. In fact the 'bodger' was for many generations a skilled craftsman. He lived and worked in a hut in a suitable beechwood, particularly in the Chiltern Hills. Here, with hand tool and foot lathe, he would rough out vast numbers of chair-legs which would be bought up by nearby furniture-makers in High Wycombe (where a famous firm still works), for the completion of the chairs.

The seats of those chairs were most likely made of elm wood, for a characteristic of this is that it does not split easily. The elm, even if it does not have such a fine crown, far outgrows the beech in height and sometimes reaches 150 feet in its hundred-year career. It usually grows in isolation although large numbers of elm trees were planted in the hedgebanks that were established during the land enclosures, early in the last century.

Because of its unreliable roots, the elm is very prone to come crashing down in high winds and cause accidents. Perhaps that is the reason for the old country-saying:

'*Ellum hateth Man, and waiteth*'.

It is appropriate therefore that one of the continuing uses for elmwood is the making of coffins. But in fact its use in this way is because of its resistance to decay. For this reason too, water-mains were made of bored elm wood from the days of the Romans, the lengths being joined by iron collars. Some of these wooden water-mains from Roman times, taken from beneath the Bank of England, can be seen in the London Museum.

Another English hardwood which has been used for more elegant purposes is the lime. Nowadays it is known chiefly through some of the splendid avenues gracing various parks, a fashion started in Restoration days. At one time, however, the lime grew prolifically as a wild tree, a fact suggested by place-names such as Linstead, Lindridge, Lindal, Lyndhurst and many others. These are derived from linden, the other name for the lime.

The making of hat blocks is the main use of limewood today. But its best claim to fame is the fact that this soft, smooth, pale-yellow, even-grained wood was used by the seventeenth century wood-carver, Grinling Gibbons.

The biggest use for the sweet chestnut (which some experts say was introduced to Britain by the Romans, while others credit the Normans) is for hop-poles and clothes-props and fencing stakes. The wood for such purposes is produced by the very ancient practice called 'coppicing', which dates back to medieval times.

Coppicing is carried out by allowing the shoots from the stump of a felled tree to continue growing in the form of long, sturdy, pole-like shoots, which are cut back every ten or twelve years.

Even today something like 300,000 acres of coppice woodlands are still maintained, mainly in southern England. In addition to the sweet chestnut the practice of coppicing is carried out with other hardwoods such as hazel, alder, hornbeam and birch.

As in many other cases, the sycamore was originally introduced into Britain as an ornamental tree. But it was soon realized that its strong, light-coloured, fine-grained, easily polished wood was eminently useful for elegant furniture which, particularly in the seventeenth century, became more fashionable than the cumbersome oak that had been popular until then.

Nowadays the sycamore is used for such purposes as kitchen tables and wooden spoons, and for the making of rollers in textile factories. It is still useful enough for nearly 90,000 acres of sycamore to be cultivated in Britain.

Another hardwood introduced in the Middle Ages was the walnut. This came from Asia Minor and was known in Anglo-Saxon times when the oil from its nuts was highly sought after for domestic purposes.

The walnut is not hardy enough to stand the climate well, so that it is only found in solitary state. Yet in Stuart days and later, it was immensely valued in furniture making, although most of the wood had to be imported from France. Nowadays the wood continues to be prized, especially in the making of high quality veneers and also for gun-stocks.

The horse-chestnut, one of the loveliest and most popular of English trees, is used only for making toys. Its timber is soft and white and it was brought to England in the days of James I.

In spring, its beautiful massed 'candles' are unrivalled by any other tree. In autumn the glossy nuts which burst from their spiked shells, are still used in the fierce game of conkers.

14 · Softwoods

The only commercially valuable conifer or softwood native to Britain is the Scots pine, which grows right across northern Europe as far as the Ural Mountains.

In its lifetime of eighty to a hundred years, it can turn into a magnificent straight-boled tree of 180 feet. Being quick-growing and hardy enough to flourish on poor soil, strong yet easily worked, it is exceedingly valuable. Railway sleepers, telegraph poles and pit-props are all made of Scots pine. When it is sawn into planks it becomes the 'red deal' which is used so extensively in house-building and carpentry.

It could be said that commercially, the twin of the Scots pine is the Norway spruce, whose timber is known as 'white deal' because of its lighter colour. This tree was introduced into Britain in the sixteenth century and is now widely used in afforestation. It is used for much the same purposes as the Scots pine, but also for paper-making, which will be described later.

A slightly later arrival in Britain was the European larch, which was introduced from the Alps in the seventeenth century. Though it never attains great girth, it is valuable because it matures in sixty or seventy years and can be used before then. Strong and durable, it is of great value for a number of purposes from boat-building to fencing.

Quick growth is also a characteristic of the Sitka spruce, which is known in the timber trade as the silver spruce. It is used for more or less the same purposes as the other softwoods, but it has the added distinction of being used in some aspects of aircraft construction.

Although the Sitka is impressive, with its ability to put on three feet of growth in a year and climb to a height of 250 feet in its native British Columbia, the grandest of all conifers introduced into Britain is the Douglas fir. On the Pacific coast of America, the Douglas can attain a height of 300 feet and a girth of 40 feet in little more than the life-span of a man. In fact, the record height attained by a Douglas fir was 417 feet. It grew in Vancouver and was the tallest tree ever known.

This tree is named after the great botanist David Douglas, who introduced it during the reign of George IV, together with the Sitka. The Douglas is as useful as it is stately and its timber is employed for many purposes from house-building to plywood-making and from pulpwood to telegraph poles.

All these and various other softwoods are now cultivated in Britain. Among them there are the ramrod-straight Corsican pine, the fast-growing Japanese larch from the slopes of Mount Fuji-yama and the Lodgepole pine, so called because American Indians used it for the framework of their lodges or wigwams.

The planting of conifers has often been objected to on aesthetic grounds. Even in the eighteenth century when the Duke of Atholl

19 Pine trees being pruned in the extensive softwood forests of New Zealand's North Island

was carrying out his afforestation programme, the larches he planted were sneeringly referred to as 'proper ornaments for a desert'.

Nevertheless, there can be no denying the immense value of these conifers not only in Britain where they have become naturalized, but in the world generally. Although there are approximately 2,500 million hectares of broad-leaved forests as against 1,280 million hectares of conifers, the latter supply 80% of the world's commercial timber.

But in any case, trees of all kinds, whether deciduous or coniferous are being planted more and more for the pleasure they give. Architects and town-planners, conscious of Man's growing isolation from Nature, plant trees around housing estates or along motorways and in shopping centres, to provide refreshing green oases in the midst of the deserts of concrete and steel.

15 · North America: Land of Trees

When in the tenth century Leif the Lucky, son of Eric the Red, 'discovered' North America, or in other words was the first European to set foot there, he named part of it Markland, which meant Land of Trees.

What those stupendous, untouched forests, haunt of wolf and moose and bear and caribou, must have been like, we can only surmise. The region between the Atlantic and the Great Plains, even between ocean and ocean, was unbroken treeland.

But once North America was 'rediscovered' five hundred years after the Vikings' unsuccessful attempt to colonize the north-east coast, the trees began to fall like ninepins. The million or so American Indians who inhabited the forests had of course used the trees for various purposes, for canoes, wigwam frames, fuel, medicine, and suchlike. However, their needs were modest compared with the greed of the white man.

Thoreau was undoubtedly right in asserting that wood to

the settlers was of 'a value more permanent than that of gold'. As we have seen, it was also of immense value to the Royal Navy, from the days of Captain George Weymouth, who in 1605 caused a sensation in English naval dockyards through the cargo of magnificent white pine that he had brought across the Atlantic. But even after trade was for the time being severed between Britain and America in the 1770's, the young United States had more than enough scope at home for its phenomenal timber.

Following the usual pattern, the American pioneer, in order to till the virgin land, had first to fell the trees. With these he built his log-cabin, while with the rest of the timber he warmed his house or made pitch and turpentine from it to protect the structure. In many places he even had to build 'corduroy' roads of logs across bogs.

The pioneers in many cases could not have survived without timber. For this was vital in the construction of blockhouses and palisades in the long and violent struggle with the Indians who naturally resented the invasion of their homeland.

Staging-posts and strong-points, such as Fort Laramie, Fort Bridger, Fort Hall and many others that were key-names in American history, were simply log-built stockades.

Not only the economy of individuals, but also that of many of the original Thirteen Colonies was dependent largely on their 'lumber'. Although Virginia is renowned for its tobacco, one of the primary reasons for colonizing it was its wealth of oak and ash and elm and walnut.

At one time timber was even made legal tender in the state of New Hampshire. Because of her abundance of trees, Maine was nicknamed the White Pine Tree State – and bears an image of that tree on its flag. Even a tragedy such as the Civil War led to many Southerners in the ruined Confederate states recouping

o *American loggers in their bunkhouse around the turn of the century* their fortunes through timber which was needed in huge quantities for the reconstruction that followed.

The world has never seen such a rapid settlement of a 'new' country in modern times. Immediately America ceased to be a colony, she expanded swiftly. In the thirty years from about 1790, her population mounted from four millions to nine and a half millions. From then on it doubled every twenty years or so.

All these new Americans had to be housed, warmed, provided with household furniture, ships and carts and sledges and countless other articles. The lumberjack became the most hard-worked figure in the country. He could scarcely keep pace with the demand for timber. It is said that between 1780 and 1930, 140 million hectares of American forestland were cut down.

With no thought as to whether the trees could keep pace, the lumberjacks went on cutting through the forests, which must have seemed inexhaustible. As the line of the forests retreated in the face of the axe, new settlements, which were soon to become prosperous towns and cities, sprang up. New farmland was established to feed the millions of emigrants who were flocking across the Atlantic.

The timber resources of Canada were exploited later than those of the United States. Even after the War of Independence cut Britain off from American supplies, lack of foresight on the part of various home governments caused Canada's lumber industry to remain virtually non-existent until the crisis of the Napoleonic Wars.

Today, Canada has nearly one million square miles of forests capable of producing usable timber. Most of this – over 80% – is publicly owned and produces annually well over 3,000 million cubic feet. This is a huge amount which is, however, considerably less than the annual growth of the forests.

Something like 150 different kinds of tree are native to Canada, but of these it is the spruce, Douglas fir, hemlock, cedar, pine, balsam fir, yellow birch, maple and poplar which are the principal commercial species. As in the case of the United States, by far the richest lumber area is on the west coast: namely, British Columbia, which accounts for 37% of Canada's lumber.

16 · Forestry in America

Because of the pattern of the original colonization, it was at first the Northern Forest which constituted America's apparent timber wealth. For the settlers had no notion of the immensity of their adopted land. This Northern Forest of white pine and black spruce covered Minnesota, part of Wisconsin, Michigan, New York State and Maine. A narrow tongue of it extends into Pennsylvania, Virginia, North Carolina and the fringes of Georgia.

But as the developing American nation began to explode in all directions, they found other huge forests of whose existence they had been unaware. Indeed, the Central Hardwood Forest proved to be the largest of all. Its oak, hickory, elm, ash, sycamore, beech, black and red gums, poplar stretched from the Atlantic right across the Mississippi and from the margin of the Northern Forest to near the Gulf of Mexico.

Yet in spite of its size, this great tract produces only one-twentieth of America's lumber, while of even smaller commercial value is the Tropical Forest which covers parts of Florida and Texas and consists largely of mangroves and mahogany.

The really spectacular forests were found in the west. The Western Forest, scattered over gigantic patches from Canada to Mexico and consisting mainly of conifers, was a rich enough discovery. But richer far was the West Coast Forest.

Ever since the 1840's, when the intrepid 'Mountain Men', braving blizzard and flood, attacks by Indians and wolves, starvation and disease, had pioneered the Oregon Trail, one of history's great highways, America had been pushing towards her last frontier on the Pacific.

The news of the incredible, untouched forests on the West Coast caused a 'green-gold' rush, surpassed only by the real gold rush that sent people flocking to California.

This forest fringing the Pacific shore is the richest timber zone of the United States – and of the world. It is not very wide, ranging from thirty to one hundred and fifty miles across. But it runs north to south for a thousand miles, from British Columbia in Canada, down through the states of Oregon and Washington and into northern California.

But it was not so much the area of this West Coast Forest that staggered the newcomers. It was its character. Nurtured by the heavy rainfall from the ocean, the trees, soaring redwoods and splendid Douglas firs, grew fast and thick. It was said that one acre of West Coast Forest contained more timber than five acres of the Northern Forest.

Men fell over themselves to ransack this packed treasure-house of Nature. Some timber firms of Maine even built their own ships to transport men and equipment round Cape Horn and up to the west coast. An enterprising timber-merchant, Frederick Weyerhaeuser, founder of perhaps the largest company in the world, bought two million acres of forestland there, when one acre cost about six dollars.

From the traditional timberlands of the north-east, special trains were chartered to bring hundreds, thousands, of lumberjacks 'who broke windows and kicked the doors of passenger coaches off their hinges, meanwhile bellowing to such effect that people wondered if a menagerie were passing'.

But when the bedlam of the 'green-gold' rush eased off, thoughtful voices were raised in doubt. It was realized how wrong was the careless assumption that the forests were so extensive they would last 'until hell froze one foot thick', in the picturesque language of the lumberjacks.

This haphazard, irresponsible approach was doing hideous damage, not only to the forests themselves but to the soil by erosion, and was also the cause of huge quantities of felled timber being wasted. Despite her immense size, America was in danger of irrevocably ruining her land.

Fortunately the realization that something must be done fast coincided with Theodore Roosevelt becoming President of the United States in 1901. He could plainly see the urgent need for the conservation of America's natural resources in general and of her forests in particular.

As far as forestry was concerned, Roosevelt's right-hand man was Gifford Pinchot, called by the Americans the father of modern forestry. After studying in Europe, Pinchot returned home and established several forestry schools and was appointed first Chief of the United States Forestry Service.

Urged on and advised by Pinchot, President Roosevelt in nine years succeeded in having 120 million acres – nearly the size of Spain – designated as national forestland and America's forests were saved, in the nick of time.

It was a policy of enlightened self-interest on the part of the Americans. They came to see that by sensible planning and conservation, it would be possible not only to safeguard the trees, but to ensure at the same time that even more timber could be extracted than before, without the danger of destroying the forests and the soil.

Pinchot would certainly have approved of the extensive tree-farm movement which is now doing so much to make Americans aware of their forest-heritage.

17 · Tropical Trees

From oak to Douglas fir, from Californian redwood to Japanese larch, from Chiltern beech to Norway spruce, the different varieties of trees grown or used commercially in the temperate zones are not large in numbers.

But as for the timbers of the tropics, a far longer list could be compiled. The names of many sound almost like magic incantations: Wawa, from Ghana; Afrormosia, from Nigeria; Andiroba and Jequitaba, from Brazil; Jelutong, from Malaysia; Keruing, from Borneo; Gurjun, from Burma; Yang, from Thailand; Podo, from Uganda; Coigue and Rauli, from Chile; Iroko, from the Cameroons; Padauk, from India . . . and a host of others such as Afzelia, Banga Wanga, Idigbo, Muninga, Missanda, Muhuhu.

The trees vary in character and use as much as in name. They range from the Brazilian 'Monkey Puzzle' or the Parana pine, which is a useful softwood, to the almost sacred Sandalwood tree which helps in the manufacture of perfumes and scented soaps.

All these and many others are among the quantities of timber continually being unloaded in western countries, to be transformed into such essential articles as office furniture, shop-counters, the decks of ships, passenger cabins or railway wagons.

Some tropical timbers have been used by western countries for many generations; indeed, in some cases for several centuries. The valuable timber trade of Borneo was started by Sir James Brooke, the 'White Rajah' of Sarawak, a hundred and thirty years ago.

He went there first, simply to convey British gratitude for help given to some shipwrecked sailors. A rebellion was in progress and Brooke took charge of the local rajah's counter-measures so efficiently that he was installed as rajah himself.

The timber industry of Burma dates back to the 1880's. With the support of the British Government, the Bombay-Burmah Trading Corporation extracted a concession from the then King Thibaw and Queen Supayalat. Rubies, silver and oil were featured in this concession, but one of the most important items was teak, among the best known of tropical timbers.

This deep reddish-brown timber is extremely heavy and tough, qualities augmented by its natural oil, which protects it from decay and exposure. This is why it has been used so much in the building of ships and docks.

Teak brings to mind the work of elephants. The natural intelligence and amenity to training of these animals, even after they have lived wild in the jungle for years, makes them invaluable in the heavy labour entailed, both in Burma and India. Teak flourishes in country where the most powerful tractors would be useless. The elephants are indispensable for lifting, hauling and pushing the giant logs to the nearest waterway, where they are roughly made up into ponderous rafts. These rafts, with a bamboo shelter for the men, are a feature of some eastern rivers.

Fig. 7 *Elephant working with timber in Burma*

In a more modest way, shire horses are still used for hauling timber in British Forestry Commission plantations in difficult country, while huge teams of oxen were once used in American and Australian forests.

Many of the best known tropical timbers have long been used for making elegant furniture. Among them are rosewood, so called for its deep red hue; also satinwood, equally well named because of its smooth, lustrous finish.

But for centuries, the most famous tropical wood has been

mahogany. Though this tree was known as far back as the late sixteenth century, it was not until the reign of Queen Anne that it began to be imported seriously into England.

Until the days of the Commonwealth, oak had been the principal wood used for the heavy furniture that prevailed. But two factors contributed to a change of fashion. The returning cavaliers brought back gayer ideas from the French court where many of them had spent their exile. Then, a few years after the Restoration, the Fire of London destroyed not only many houses, but the furniture in them as well.

When supplies of walnut became exhausted, mahogany entered into a long period of ascendancy in the making of furniture. It was said that mahogany 'of all the woods is best suited to furniture where strength is demanded. It works up easily, has a beautiful figure and polishes so well that it is an asset to any room'.

Those words were written by Sheraton, the eighteenth-century cabinet-maker, whose name, together with the names of Hepplewhite and Chippendale, represents a period of English craftsmanship which has never been equalled. Mahogany still comes to Britain from tropical countries on both sides of the Atlantic, such as Honduras and the Congo. One of its uses nowadays is in the making of television boxes.

18 · Developing Countries, Developing Forests

The use of tropical timbers constantly increases, both because more timber is needed in general, and because the developing countries realize it is potentially an immensely valuable part of their economy.

As yet the timber resources of such countries are far from being exploited to anything like their full extent. Inventories carried out by the Food and Agriculture Organization (FAO) of the United Nations, show that although nearly 30% of the world's land surface is covered by forest, only one-third is being utilized.

They show also that coniferous forests of the northern hemisphere comprise 90% of the total softwood resources and provide more than half the world's exploited timberland. Because of longer experience and better techniques in reafforestation and logging methods, the consumption of timber in this great northern belt is balanced by the annual growth.

But as far as the developing countries are concerned, much remains to be done, both in reafforestation and commercial use. For example, one-fifth of Asia is afforested, compared with more than one-third of North America and the Soviet Union.

Even the African countries, which export timber to Britain, still have to import timber from the better organised forests of Europe. In spite of the immense hardwood resources of the Amazon and the Orinoco, some South American countries still find it more economical to import Douglas fir from Canada and America, four or five thousand miles away. Likewise, an African country such as Sudan exports large amounts of gum arabic, product of an acacia tree, but has to spend the foreign currency she thus earns in importing softwoods for building purposes.

Altogether, the underdeveloped countries have a trade deficit in forest products amounting to more than 300 million dollars, yet Latin America, Africa and Asia probably have between them more than 2,300 million hectares of forest.

But various agencies are making immense efforts to help these developing countries. Typical is the Commonwealth Forestry Institute, attached to the University of Oxford. Its activities include the teaching of forestry methods for an honours degree, the providing of certain services such as botanical identification and statistical information, and the investigation of diseases and pests. One of its most useful services is the scrutinizing of all forestry publications throughout the world. In this it covers some forty languages and issues abstracts of the most important material. Its library has the finest representation of world forestry literature available.

But the biggest work in the field is provided by the FAO which is progressively helping the developing countries to improve their forestry techniques. They have provided money and experts to set up Forestry Training Establishments, Research Institutes, University Schools in countries as far apart as Chile and Malaysia. They have encouraged the spread of coniferous trees because of their quick growth, eucalyptuses because of their usefulness against erosion and for local fuel.

They have helped in setting up sawmills and pulpwood factories, provided botanists and wood technologists. They have shown the economic usefulness of forests in protecting wildlife, which is steadily becoming an important source of revenue as a tourist attraction.

Though a description of the work of the FAO would require a separate book, it is appropriate to mention some of its little publicized, but exciting and valuable projects. In co-operation with the Indian Government it is battling to defeat the notorious Rajasthan dust-bowl (whose debris sometimes chokes the houses and offices of Dehli seventy miles away). Here, along more than 4,000 miles of a special irrigation canal system, it is planting 10 million trees in the form of shelter-belts, wood-plantations and wind-breaks, all an essential contribution to the work of soil-conservation and irrigation.

In Greece it is helping to rehabilitate 12,000 hectares of land denuded of its trees by over-pasturing of goats. In a four-year programme in Algeria, the FAO is helping to plant 42 million trees to re-afforest land partly devastated by napalm bombing during the war of liberation against the French. It is planned to recover more land by tree-planting. Not only do such schemes give vital help in raising the standard of living, but they also provide employment for thousands of people and at the same time bring hope and self-respect for the poverty-stricken families concerned.

Much of the FAO's work is helping to find the best kind of

21 Plantation of Aleppo pines and eucalyptus in Algeria; part of the FAO programme

trees for the soil and climate in question. A significant project in this respect is the tree nursery in the Zerniza forest of Tunisia. Here one hundred acres of specially planted trees – eucalyptus, pine, cypress, casuarina – have been laid out like an immense outdoor laboratory in which every possible detail is noted in relation to growth, reaction to conditions, improvement by grafting and so on.

The scientists in charge even 'listen-in' to the workings of the trees as they grow, by means of a dendrograph, an ingenious device fixed to a tree and connected to recording apparatus which registers on a graph the slightest change in the living processes of the tree.

One of the most urgent problems in the underdeveloped countries is the matter of communications and accessibility. Whereas something like 98% of European forestland is accessible (the Russians claim 100% of theirs), only about 32% of South and Central American forests, 50% of African, 63% of Asian and 27% of the Pacific area are accessible.

These tropical forests were long known only to primitive tribes and hardy explorers. The World Health Organization, by its campaigns against such scourges as malaria and yellow fever, has contributed greatly to the possibility of opening up the jungles. With the help of the FAO, countries such as Ghana, Nigeria and Thailand have increased accessibility to their forests by 20%, including roads, river transport and railways. To return to Chapter One, it is quite possible that in the future large hovercraft, balloons, helicopters will be used to extract timber from impenetrable jungle.

The jungles of Latin America alone contain vast natural reserves of timber – almost one-quarter of the world total – which are as yet largely untouched because of this question of access. However, as technological advances are made, it is likely that such tropical forests will not only help immensely in raising the living-standards of the people concerned, but will also contribute to the ever-growing demand for timber throughout the world.

19 · Timber in Australasia

Only part of Australia lies within the tropics, and her forests range from the native tropical or sub-tropical rain-forests on the coast of New South Wales and Queensland, to man-made plantations in the temperate southern regions, which include Tasmania.

As in the case of Canada, it was the Napoleonic Wars that gave rise to the Australian timber industry. Timber had become so literally a matter of life and death for the British Navy and therefore to the British nation, that the government was forced to seek supplies in the farthest corners of the world. In 1804 a few cargoes, mainly of red cedar and blue gum, were shipped to British naval dockyards. It was an immense voyage, taking many months, but is a vivid illustration of how precious every stick of timber was.

From that crisis start, the Australian timber industry has expanded until now a hundred thousand people are employed in all its branches – in forest services, in harvesting forest products, sawmilling, and in factories dependent on forest products for raw materials.

Productive forestlands occupy some 63 million acres, an area roughly equivalent to the whole of the United Kingdom. But out of this, 25 million acres produce only fuel-wood, while the total represents no more than 3% of Australia as a whole, though it has to be remembered that almost one-third of the continent consists of desert or barren land. In fact, recently the output of timber has shown signs of declining, because of the over-exploitation of the native rain-forests in the past and the fact that new, man-made forests have not yet had time to fill the gap.

Rain-forest or jungle cannot easily be replaced by artificial planting and depends largely on natural regeneration, which is a slow process. So, as with many other countries trying to restore or increase their timber reserves, Australia's reafforestation programme is mainly concentrated on conifers from California and Mexico and even the West Indies. Twenty-five thousand acres of these and other softwoods are being planted every year to augment the 600,000 acres of softwood plantations already established.

But there are still ten times that area of Australian conifers, such as the hoop, bunya and kauri pines. Simultaneously, the Australians are expanding their forestry services, chiefly through the State Forestry and Timber Bureau. This maintains a Forestry Research Institute and also a Forestry School, which works closely with the universities. As with the world in general, it is realized that forestry and its kindred subjects have become a science requiring the best brains.

There is certainly immense scope in Australia's trees for her future needs and development of trade. They include several which are well-known in furniture-making, such as coachwood, black in colour and sweet in scent. Equally useful for panelling and floorboards is the beautiful chocolate-coloured black bean. The Queensland maple (so called because of its likeness to real maple) provides much of the wood used by British Rail for its carriages.

Sandalwood is exported to China for the manufacture of joss-sticks. Turpentine wood goes to Britain for making dock-piles, as it is resistant to the pernicious teredo worm that does so much damage to less durable timber. Ironbark has supplied the timber for many a vessel operating in the demanding sub-arctic seas. Many of the world's railways run over sleepers made of jarrah timber.

Crow's ash, brigalow, gidgee, celery top pine – there is even one called raspberry jam because of its peculiar scent, but the best known native trees are the eucalyptuses. Both in size and in number these are the most dominant species in the country.

To many people the pungent scent of eucalyptus makes them think of cough cures. But eucalypts provide timbers which are of considerable value because of their strength and resistance to decay and also to termites, always a threat in tropical countries.

The name eucalyptus comes from the Greek word meaning 'well covered', for the flowers are protected by a sort of cap before they blossom. But though they have this characteristic in common, the eucalypts vary widely otherwise, for some 500 kinds exist. One group of them alone, the bloodwoods, so called because of the red gum they exude, comprises more than thirty different trees, while other groups include the tallow-wood, flooded gum, red 'mahogany' (not real mahogany), jarrah, tuart, and many others prized for their timber or their oil.

One of the most magnificent trees in the world is a eucalyptus: the karri of Western Australia, hardly surpassed even by the giant

trees of Pacific America. But the most famous of all is the blue gum. This has been successfully naturalized in countries such as Israel and Ethiopia for soil-preservation purposes, while in parts of India it has been planted especially to supply firewood in cases where there is no fuel other than dried cow-dung. More important, the eucalyptus has proved of immense value in countries threatened by the encroachment of the Sahara Desert: Morocco, Tunisia and Algeria. Tiny seedlings planted at the edge of the desert grew to a height of twelve feet in four years, while in one area, according to the Moroccan Forestry Commission, a eucalyptus plantation doubled the rainfall of the neighbouring desertlands.

New Zealand originally had a larger proportion of forestland than most other countries. Out of her total area of 104,000 square miles, perhaps two-thirds were densely covered by forest until the early nineteenth century.

During their six or seven centuries' occupation of the country, the Maoris treated the forests rather as the American Indians treated theirs. They used them for shelter, food, hunting, and occasionally burned down patches in order to clear and till the soil in a primitive way.

But when the white man came, once again the trees began to tumble. In a crescendo of activity throughout the last century, particularly the second half, millions of acres of forestland were destroyed by axe and fire.

Of course, much of the timber the new immigrants cleared was needed for the development of the country. But much of it was wantonly wasted. It has been estimated that for every acre from which timber was felled and properly utilized, at least five acres were felled and simply burned on the spot.

Felling eucalyptus in a log camp near Melbourne, Australia

The land thus cleared was later developed into some of the richest dairy farmland in the world, but in many regions of New Zealand the thoughtless clearance of the forests led to large-scale erosion and flooding.

At last, and like the Americans, almost too late, New Zealanders realized their forest inheritance was vanishing. Like the Australians, too, they found it was in most cases useless to rely on natural regeneration of their native trees, which consisted both of conifers and deciduous kinds. Fine though these were, they were too slow-growing and Nature will not be hurried, though she will often co-operate.

As a result, New Zealand is in the gradual process of completely changing the character of her forests. In the future, the chief purpose of her natural forests will be not the production of timber but the equally vital purpose of soil-protection and water control.

For commercial timber afforestation, New Zealand, again, has gone to other countries. Douglas fir, ponderosa pine, larch and many other 'exotic' or foreign trees are being planted. But the most important newcomer is the radiata pine, whose prolific seeds make it ideal for natural regeneration as well as for aerial sowing, so that it now provides 90% of New Zealand's sawn timber.

It was with this tree that the New Zealand Forest Service created the largest man-made forest on earth: the two hundred thousand acre Kaingaroa State Forest in North Island.

Aerial view of clear felled settings in part of the Kaingaroa Forest, w Zealand

20 · The Red Sash Brigade

Inevitably, the trees that are grown commercially for timber have eventually to be felled. Lumberjacks exist wherever such work has to be done, from the jungles of Burma to the taiga of Siberia. But undoubtedly the classic lumberjack of tradition hales from America and Canada.

Long ago these 'loggers' earned themselves the nickname of the Red Sash Brigade because of their delight in bright-hued clothes. The lumberjacks have retained this predilection for red, particularly in the way of shirts and mackinaws, the short, heavy jackets of woollen plaid which are as much part of their 'uniform' as their steel-calked boots.

For months on end, including the worst of the northern winter, their only company the wolves, they would live in the sombre isolation of the forest. They were housed in primitive shanties with roofs of bark, walls of moss-stuffed logs and triple-decker bunks whose bedding consisted of a truss of hay or a few branches.

Apart from the bunks, the only furniture was a table and a couple of benches made from a tree split lengthwise. Their food consisted of salt pork, beans, rye-bread and molasses, and tea which was reputed to be strong enough to float an axe on.

But the living conditions of the lumberjacks were mild compared with the dangers present in every aspect of their work. Until the 1880's, lumberjacks used only the axe for felling – and these axes were said to be sharpened so devotedly that it was possible to shave with them. A good lumberjack could fell a tree so accurately that when it crashed down it would drive into the ground a stake previously set up in line with its anticipated fall.

But many a man, ignoring the traditional warning cry of 'Timber!', was also flattened by a falling tree, whose branches might bludgeon him nearly a hundred yards away, while the 'kick-back' of its butt, the sawn end of the trunk, was another hazard.

A further danger was in driving the huge horse-drawn sleigh-loads of timber down to the nearest river, the main thoroughfare for the logs to reach the sawmills. Some of those sleighs were built to take timber sixteen feet across and totalling 60,000 pounds in weight.

But a mere average load could mean mortal peril if it got out of control going down an ice-coated track with the slightest gradient. To provide a good, smooth, fast surface, a man in a sled would sometimes be employed to sprinkle water over the track during the night, so that it froze solid in a glassy film.

If the teamster or driver failed to brake his sleigh with chains, even on the least incline the heavy load might begin to slither down faster than the team of horses could gallop. Eventually the wretched horses would be overwhelmed by the gigantic load of logs that crashed down on them.

Such a calamity was known as being sluiced and was a fearful disgrace for a lumberjack – if he survived to tell the tale.

But undoubtedly the greatest danger entailed in the lumberjack's work was the Spring Drive. The felled timber was piled on to the frozen rivers, where it lay until the thaw released the pent-up waters weeks later.

When it did, the result was spectacular. In a thundering, booming, constant roar, the 'river-drive' began, with tens of thousands of logs sweeping downstream. Along the bank or the nearest track would gallop horses drawing the manager's cart and the supply wagons.

At strategic bends lumberjacks waited with their 'peaveys' to keep the logs on the move. The peavey was named after its inventor, Joseph Peavey, a nineteenth-century Maine blacksmith. It is still the lumberjack's most valuable implement after the axe and the saw, consisting of a very strong hickory pole some five feet in length, armed with a heavy, sharp steel spike and a barbed, hinged hook. It revolutionized the river-drive, especially in dealing with the inevitable log-jams.

A log-jam could reduce the flow of water in a sizeable river to a trickle. Villages could be flooded if it were not cleared in time. Perhaps the worst on record happened in 1886 on the St Croix River in Minnesota. Many miles of logs were jammed, comprising 150 million feet of timber. It took two hundred lumberjacks, a hundred horses and a couple of steamboats seven weeks to clear the blockage.

It was in a log-jam that the real white-water men, as they were proudly known, came into their own. Dancing, balancing, acrobating in their steel-spiked boots, they would work desperately among the heaving mass of timber, trying to free the king-log at the centre of the jam.

24 Ox team dragging fir logs to a railroad spu[r] in the early days of American logging

25 *Canadian 'boom man' in his small scooter boat, rounding up logs (British Columbia)*

26 *Aerial view of 'tear-drop' log booms in Ontario*

Once that key log was located and the jam began to break up, the lumberjacks had to move fast. To lose one's footing amid that maelstrom of wood and water meant certain death.

A man's only memorial would be his name carved on a neighbouring tree, while if his body was recovered, his mates would hang up his boots on a branch in tribute.

The river-drives still go on, not only in North America but in Scandinavia and Russia as well, and log-jams still occur. Occasionally, in extreme cases, resort has to be made to dynamite. Where rapids are too shallow to allow the passage of logs, a by-pass in the form of a timber-slide may have to be constructed. In cases where lakes form part of the river-drive and the logs are becalmed or driven ashore, an amphibious craft called the alligator is used. This can cross intervening land between one lake and another and shepherd the logs on into the main riverway, so that the drive can continue.

Obviously the river-drive can only take place on suitable waterways or floatways as they are sometimes called. Where necessary, on large lake or river-mouth or sea-coast, the logs are mustered into huge rafts or booms which are towed to the sawmill by powerful tugs.

21 · Today's Lumberjacks

Nowadays, before ever a lumberjack starts up his mechanical saw, an accurate assessment of the forest's potential is made. Aerial photography is sometimes used to estimate the quantity of timber available. Photographic intelligence was developed to a state of pinpoint accuracy during World War II to identify enemy preparations. Now it is being used to count trees and generally survey and assess them.

Even so, the main work of estimating a forest's potential is carried out on foot by the timber-cruiser, as he is known. He has an important job in modern planned forestry, whether in the modest plantations of the British Forestry Commission or the giant treelands of Canada and America and other countries.

The timber-cruiser not only counts the trees in selected areas so that a detailed one-fortieth sample of the whole forest is obtained, he also records the various species of trees involved as well as their age and state.

The preliminaries are much simpler where clear-cutting is to take place, for this means that an entire forest or block is to be felled. Sometimes, when trees are of different ages or quality, selective cutting is decided on. In this case the trees to be felled are marked by means of a special paint-gun. In certain cases seed-tree cutting is employed: in this case a few trees are left in every acre of forest, to act as seed-trees until the subsequent new seedlings become established.

In addition to knowing the likely quantity and quality of timber to be extracted, it is necessary to obtain an accurate idea of the terrain. The character of valleys and ridges, the contour-lines, the gradients, swamps and rock formations must all be known.

This is essential in order to decide on the most favourable situations for the siting of skidways and roads, log assembly-points and camps and all the extensive operations connected with the extraction of the timber.

In Canada, horse-sleighs are still used for initially transporting loads of timber. But the rolled, glazed ice-roads along which they operate are heavily sanded at strategic gradients to help in braking. Still in use, from Russia to India, are the old corduroy-roads. The 'skid-road', as the corduroy-road is called in America, consists of felled, trimmed logs set parallel to each other at close intervals. In swampy land it was indispensable for the huge teams of oxen which were formerly used for hauling timber and it is often still needed even today.

The importance of access to the forests by ordinary roads has grown correspondingly both with the increased demand for timber and the possibility of using heavy machinery and equipment, such as caterpillar tractors, grapple-loaders, or huge diesel trucks. For obvious economic reasons, the lumbermen aim always

27 and 28 Hauling and loading logs on site

to reduce the hauling distance from stump (that is, the cutting site) to road as much as possible. In a typical, well-organized forest region such as South Sweden, for example, the average is half a kilometre – nearly 550 yards.

In Britain, though forestry roads are equally essential, they are naturally not extensive. But in the Soviet Union there are more than 55,000 miles of logging roads of different types. Special logging railways have been built there, too, as is the case in countries such as New Zealand and India, together with all the attendant engineering essentials such as bridges, viaducts and embankments.

Long stretches of first-class motor-roads, along which their employees often commute to and from work every day, have been built in America and Canada by some of the large forestry companies. Any one of these may employ as many as 15,000 men in their forests at the height of the logging season. In lumbering, something is going on all the year round, but most of the felling, at least in the northern hemisphere, is carried out in the autumn. This is because the sap is less active and the cut timber is therefore less susceptible to rot. In timber countries such as Canada the forester regularly inspects and weighs 'humidity sticks' kept in a special gauge, to determine if the woods are too dry for logging operations to commence.

The camps the lumberjacks live in have changed, too. Of course, on the logging site itself life is still tough and arduous and dangerous. But the men – lumberjacks, mechanics, drivers, explosive experts – enjoy a far better standard of amenities than the Red Sash Brigade could ever have imagined. In Britain, special forestry community villages have been built for Forestry Commission workers and their families, complete with shops, post offices, schools, playing-fields.

A steam-powered donkey engine, once used for dragging logs
American loading sites. Here it is being transported to a new location

30 (above) New Zealand diesel locomotive hauling a logging train

31 (below) Working with a power saw

32 (opposite) Logs are pulled by cable to this portable steel tower for loading on to trucks, Washington, U.S.A.

BERGER
842

НЕ КАНТОВАТЬ

But the biggest changes have come in the felling and extraction of the trees themselves. A quarter of a century ago, many lumberjacks were still cutting trees with axe and cross-cut saw. Nowadays the gas-powered saw, one of the latest developments in logging, enables one man to do in five minutes work which previously took two men at least half an hour to do. The saw the lumberjack uses is always his own property and responsibility.

As for the extraction of the timber, increasingly, where the terrain is suitable, immense sky-cable systems, like giant ski-lifts, are used to transport loads to assembly areas. Even greater advances are planned: in America and the Soviet Union, experiments with mammoth balloons for transporting logs have been made, while it becomes highly possible that eventually helicopters will be used to carry timber direct from forest to sawmill.

Inflating a balloon in a Russian forest. It takes 700 cubic metres of gas

22 · The Saw: from Flint to Steam

The principle of the saw has been known for thousands of years, since Neolithic Man fashioned crude saws out of flint. Mechanical saws, powered by wind, water or treadmill, were first experimented with in the thirteenth century. There were even elaborate but slow-moving vertical three-bladed frame-saws. Not surprisingly, the Dutch extensively developed the use of windmills for driving saws. By the sixteenth century the French were using waterwheels which, by means of simple reciprocal machinery – that is, with an alternating backward and forward movement – drove the saw up and down.

An enterprising Dutchman tried to set up a wind-powered sawmill in England during the reign of Charles II, but it was wrecked by the local hand sawyers, fearful of redundancy.

In view of all her technological contributions to the Industrial Revolution, it is surprising that England was so laggard in adopting new sawmilling methods. But even in the mid-nineteenth century, half the timber used by the Royal Navy was still hand sawn.

One of the commonest methods was pit-sawing, used as far back as Roman times and which can still be seen in India and Africa. In this, the tree-trunk to be sawn was marked off for planking and straddled across a rough scaffolding erected in a pit sometimes fifty feet long. It had to be deep enough for the 'pit-man' or bottom sawyer to work underneath – where his eyes and lungs were filled with sawdust.

Meanwhile, the top-sawyer controlled the long saw (up to ten feet in length if two pitmen were employed) as he stood on the tree. Sometimes a whole series of a dozen or even a score of parallel 'kerfs' or cuts were made and kept open by wedges, before the tree was moved forward by means of log-rollers levered by crow-bars.

The most revolutionary advance in sawmilling was the development of the circular saw. The old reciprocal-action saws were slow and cumbersome and it was said that a sawyer could sit on a log and eat his dinner at leisure before he had to move. The circular saw, on the other hand, was capable of continuous action.

It was a Southampton carpenter, Walter Taylor, who is credited with first introducing the circular saw into England. To begin with he used a horse as the motive power, but then, prospering through the demands of the Navy, he took over an old cornmill, Woodmill on the River Itchen.

However, by far the most important development was the harnessing of the steam-engine to the circular saw, chiefly by Sir Marc Isambard Brunel. Encouraged by Sir Samuel Bentham, significantly Inspector-General of Naval Works and himself a pioneer in sawmilling techniques, Brunel developed the process so effectively that by 1808 ten unskilled men were able to do the work of a hundred and ten skilled artisans. It is the circular saw

34 Saw repair shop in Waipa State Forest sawmill, New Zealand

which is most commonly used in Australia today.

In contrast with England, America had established a number of sawmills even in early colonial times, in spite of the disapproval of the home government who did not want her colonies to become too independent economically. These vertical, reciprocal frame-saws or gang-saws were driven mainly by wind or river power. Sometimes the tides were used.

Some fifteen years after the War of Independence, American methods had improved to such an extent that a man and a boy could saw four thousand feet of timber into planks of up to twenty feet long and an inch thick in a day's work.

A quarter of a century after Walter Taylor began working in Hampshire, a New York blacksmith, Benjamin Cummings, made and patented a circular saw, the use of which eventually spread rapidly throughout America. At the same time frame-saws with up to thirty vertical blades were also developed.

By the time of the Civil War there were more than twenty thousand sawmills in the United States, ten times the number existing half a century earlier. America was losing no time catching up with the position in the Baltic countries, whose coasts had for long been dotted with hundreds of sawmills. Even a poorly forested country such as Holland had become so advanced in sawmilling that Britain often had to send timber there for sawing.

23 · Modern Sawmilling

In the days of the pit-sawyers the heavy tree-trunks had to be pushed and rolled and laboriously coaxed into position over the saw-pit. A selected tree would be 'lined out' for sawing simply by rubbing a cord with chalk or charcoal and snapping it along the centre of the top surface. The rest was a matter for the top-sawyer's eye.

But even the eye of an expert could not always tell whether the tree was foxy-hearted or rotten inside, while if an ugly shriek announced that the saw had encountered a piece of metal or stone embedded in the timber, half a day might be wasted repairing the damage and resetting the saw.

In contrast, little exists in a modern sawmill that is not automated. Even the millpond where the reserve of logs waits is steam-heated so that winter-ice does not interrupt the work. A huge, moving jack-ladder, armed with holding spikes and resembling an escalator, brings the logs into the mill. In some modern

35 A giant hammerhead crane unloads 30 tons of logs at one lift

sawmills, the 'electric elephant' with hydraulic arms can grab and lift forty tons of timber at a time.

As for the danger of metal being hidden inside the wood, a tree can be tested electronically if necessary. In the important task, too, of detecting faulty timber, an ultrasonic device is sometimes used. The ultra-sound passes much more slowly through suspect areas of timber, where the density is less. The time taken by the sound-pulses to travel through the tree-trunk is counted electronically.

Meanwhile, the operator, who is equipped with an ultrasound receiver, consults a table of standard ultrasonic velocities for various kinds of wood. By comparison, he is able to estimate the extent of rotting and the size of the affected area.

But even the latest invention is based on previous attempts or experiments. As far as modern sawmilling is concerned, the principle of the old, simple, slow-working machinery has been developed into the modern electrically powered frame-saw.

In this, the sawyer directs the entire operation sitting in a metal chair above the end of the log-carriage. As a particular log is delivered, he switches a lever to make a pair of rotating 'dogs' or grips seize the butt end and lift it on to the carriage.

Three other dogs, resembling huge pincers, keep the log in position. After the sawyer has adjusted the position of the log to the best advantage, depending on its character, a touch on another lever enables him to run the log forward on the moving carriage until its end is between rollers which feed it into the frame-saw at a speed of about thirty feet a minute.

The multiple saws can be adjusted to various thicknesses of cut as required. The usual practice is for the log to be sawn into a central 'flitch'. This is a large baulk of timber with several smaller planks, according to the number of saws, at the sides.

These side boards are next carried on a conveyor belt of moving rollers to an edging machine which trims them, while the central flitch has to pass through a second frame-saw where it is cut up into more planks. On a fast modern frame-saw, the single sawyer or operator can cut up to four or five hundred logs into planks in a working-day at a rate of less than two minutes a log.

The frame-saw is chiefly favoured in Scandinavia and the Soviet Union. It was another kind of saw that the Americans and Canadians – and also nowadays the Australians – took to more enthusiastically. This was the bandsaw and it is now predominant in the sawmilling industry. It can handle bigger stuff than the frame-saw, particularly at the outset of sawing operations.

The bandsaw was invented by William Newberry of London in 1808. In principle it consisted of a steel ribbon saw revolving

36 Log bandsaw in operation at a Devonshire sawmill

37 (above) Rolling table circular saw

38 (below) Portable bandsaw used on a forest site, Devon

round two wheels placed vertically one above the other, though horizontal bandsaws have also been used.

The Americans in particular realized its possibilities, especially after it was demonstrated at an exhibition at Philadelphia in 1876, which could be said to have started the hitherto neglected Newberry bandsaw on its career.

Nowadays, the bandsaw, equipped with similar massive lifting and adjusting devices as the frame-saw, is used all over the world. Its sometimes twelve-inch-wide ribbon works at great speed, cutting through three feet of timber a second.

The frame-saw and the bandsaw (and also the circular saw, singly or in vertical pairs) form the headrig of the modern sawmill, in other words they carry out the heavy initial work. But when this has been done, the timber has to pass through many other re-sawing processes, for which smaller bandsaws and circular saws are used, with all the aids of modern automation.

Finally, the resultant planks have to be seasoned, either by artificial kiln-drying or by natural air-drying.

24 · Paper from Timber

Nowadays one of the most important products of the forests is paper. Once Johann Gutenberg of Mainz in Germany had developed the printing-press in the fifteenth century, a growing demand for paper was inevitable. But for several centuries paper-making continued to be based on ground-up linen rags and only comparatively small quantities could be produced.

But the demand swelled progressively, especially as education spread and more people became literate. It was another German, Friedrich Keller, who in 1873 suddenly provided the answer. He developed the process of manufacturing paper from wood-pulp and a mass-supply became possible.

The union of the printing-press and unlimited, cheap supplies of paper had been long delayed, but it was to have a profound effect on people's lives, through the media of more easily produced books and newspapers with large circulations.

By far the greater part produced today is 'newsprint', on which newspapers, magazines and many books are printed. Most of this newsprint, perhaps 85% of it, is made from what is known as groundwood pulp, a comparatively cheap and purely mechanical process. It is also known therefore as 'mechanical' wood pulp. The various chemical processes produce finer quality paper but are more costly. Increasingly, groundwood and chemical pulp are being used together to make certain kinds of newsprint.

The United States and Canada are the biggest pulp producers in the world. Sweden is the biggest exporter and sends abroad 55% of her annual production of six and a half million tons.

Spruce and pine are by far the most important trees used in making paper. The timber is supplied in handy four-foot lengths and a pulp-mill's stock-pile is constantly being replenished by giant overhead stackers.

Conveyor belts carry endless masses of logs to huge revolving machines called barking drums, where the first stage of paper-making takes place. This consists of the logs being subjected to water jetted on them at very high pressure, while simultaneously they are churned round violently, crashing against each other and against immensely strong bars inside the drum.

At the end of this bruising process, the logs have been completely skinned. The next stage in the 'mechanical' process is for the purpose of breaking up the logs into the separate fibres of which the wood is composed. Forced on by hydraulic rams, or passed along mechanical tracks, the logs are fed between powerful grindstones revolving at great speed produced by 4,000 horse-power motors. Water is constantly played on to the grindstones, both to keep these cool and also to send the ground-up wood on to the various careful sieving processes through which it has to pass.

Chemical wood-pulp consists mainly of two types. The first is sulphite pulp, for which spruce is used because it reacts more favourably to this process. The second is sulphate pulp, for which the more resinous pine is used. This produces a stronger paper for wrapping purposes especially and is known commercially as 'kraft', the German word for strong, as it was in Germany that the process was first developed.

Wood used for making chemical wood-pulp goes through the same debarking procedure as mechanical wood-pulp. But instead of being ground up, the wood passes through a machine which chops the logs into small, flaky chips, an inch or so long.

The chips must next be digested. This is done by a giant boiler, forty feet high and actually called a 'digester'. It has a mammoth appetite and will take twenty tons of chipped wood at a time. To help it in its work of digestion, the chemical solution is added and the whole mixture is boiled up by means of steam pressure.

This seething brew dissolves all the lignin which cements the wood fibres together. At the end of the process, which may take up to twenty-four hours, the wood-pulp is sieved very carefully to get rid of all the unwanted chemicals and other substances. Then the pulp is mixed with clean water and sieved yet again. It should now be free of anything but the essential fibres and, in conjunction with the water, is transformed into a thick slurry, which is also what the mechanical pulp becomes.

One great advantage of chemical pulping is that the basic cellulose fibres of the wood are far less damaged than in the mechanical process. The cruder process of mechanical pulping produces bundles of short and ragged fibres, whereas the chemical method separates the fibres more gradually and individually, leaving them longer and therefore stronger.

In the case of some very big manufacturers, the pulp-mill may stand on the same site as the paper-mill. When this is so, the semi-fluid pulp, which has a consistency something like that of porridge, is pumped straight through to the paper mill as required.

More often than not, however, the mills are separate concerns. For the purpose of transport, often overseas, both mechanical and chemical pulps are temporarily converted into fairly thick, partially dried sheets. This is done mainly by passing the pulp over a belt of fine wire mesh. In this way it is dried out sufficiently for it to be cut into sheets which are then compressed into bales of about four hundredweight each.

When the pulp arrives at the paper-mill, it has to be reconverted into a liquid state. This is done by a machine called a hydrapulper. If groundwood and chemical pulp are used together, they are mixed in a proportion of five parts of groundwood to one of chemical pulp.

But in any case the pulp is broken up again and mixed in the hydrapulper at first simply with water, all violently swirled round by the revolving knives in the huge trough of the machine.

When the resulting slush or 'stock', as it is known in the paper-mill, is of the right consistency, kaolin or china clay is added. This helps to make the eventual paper opaque and improve its printing surface.

If necessary, dyes may be added to the stock to adjust the tint of the paper. Alum, a mineral salt, is also part of the recipe, for its astringent qualities have a binding effect.

After all this the stock has to be carefully refined once again. The refiner is a cone-shaped machine containing a second revolving cone fitted with blades. In this the final disintegration of the pulp into its constituent fibres is achieved.

Before the stock starts actually being made into paper, more water is added, sometimes in a proportion of two hundred parts to one. The result is a very watery mixture which obviously has to be elaborately dried out.

Stages in paper-making: (*above*) 39 *Bales of wood pulp are reduced to slush in a hydrapulper*

(*top right*) 40 *Web of paper passing through drying cylinders*

(*bottom right*) 41 *New paper being wound on to a spindle, Bowater Corporation*

FIRE

A modern paper-making machine, often called a Fourdrinier from the name of the original makers, may be four hundred feet long. The first section consists of an extremely fine wire 'cloth' in the form of an endless belt some twenty feet wide and eighty feet long, moving at two thousand feet a minute. This belt is made of phosphor bronze and brass wires in such a delicate pattern that there are three thousand meshes to every square inch of it.

By means of this sieve much of the water, many tons of it, is removed. What is left is a wet web of pulp fibres matted together. Now another endless belt made of woollen felt takes over and helps to a lesser extent with the drying-out. But it feeds the pulp web into a monster mangle. The rollers of this continue the process of getting rid of the water, while at the same time making the web firmer.

Next, the web, which can now be called paper, has to be passed through a drying process. This is carried out by hollow, steam-heated, rotating rollers or drying cylinders about five feet in diameter. There may be as many as fifty such cylinders on a modern paper-making machine and by the time they have finished their work less than 10% of water is left in the paper.

Last of all is a kind of ironing process in which heat and weight combine. For this a calender, consisting of a series of enormous and extremely heavy rollers, is used. From the calender the finished paper is wound off into rolls four miles long and up to twenty-five feet wide.

25 · New Uses for Trees

Paper is not by any means the only material obtained from processing wood. The basis of another important product is obtained from the same process as chemical wood-pulp for paper. This is rayon, or artificial silk, used extensively in modern dress-making. The original chemical pulp is further treated chemically and then pressurized through a very fine mesh. Almost as if thousands of artificial silk-worms were in action, this produces extremely fine threads, which are plaited together to make the finished substance.

Rayon is based on the tree's sugary content or cellulose, which possesses considerable nutritional properties. As suggested in Chapter Two, during times of famine in the past, men sometimes used the inner bark of trees to eke out their flour. They had the right idea, for because of this sugar-content of the wood it has proved possible to manufacture a synthetic food from it.

Twenty years ago American scientists developed a valuable

stockfeed in this way, the use of which at the time was only abandoned because the price of real molasses fell sharply. Nowadays, however, the process is again being exploited, not only as food for cattle and pigs, but also for human beings. Perhaps like the unexploited plankton of the sea, wood will eventually contribute to the food supplies of the world.

Nor are synthetic foods or sugars the only by-products of wood and the chemicals that are sometimes used with it. In the United States alone, one hundred million dollars' worth annually of drying oils, alkyd resins, soaps, lubricants, emulsifiers, are all being manufactured from the residue of certain paper-making, in other words they are largely based on processed wood.

There are many other ways in which processed wood is used, 'wood flour' being one of them, though in spite of its name it is not a food. It is obtained by grinding to a very fine consistency wood waste such as sawdust and shavings. These pass through various re-grinding and re-sieving stages and the finished wood flour is used, among other things, as a filler in the manufacture of linoleum and also in the making of plastics.

Again, chipboard is made from coarser waste, and also from specially chopped wood, which is afterwards mixed with special glues and compressed and shaped into sheets and blocks. Different processes in the way of breaking wood into its constituent fibres also produce various appropriately named fibre-boards and hardboards, in constant use for partitions or door-backings and similar purposes.

These 'made-up' woods are very light but also very strong and strength in wood is obviously of the utmost importance. Veneers and plywoods come to mind in this connection. The making of veneers was practised by the Ancient Egyptians and much used in subsequent ages by English furniture-makers.

42 Working with plywood at a Canadian mill

Its purpose was actually two-fold, not only to impart strength but to utilize as much as possible of any rare wood with a handsome grain. By cutting the wood into thin sheets and gluing it on to a less valuable wood, the furniture-makers were able to make the fine grain go much farther. But in addition they could obtain great strength by gluing sheets of veneer together. Thin individually, these sheets were nevertheless very strong when laid one on top of another with the grain of each piece at right angles to that of its fellows.

It is by exactly the same principle that the much used modern plywood is made. But whereas veneers were cut with specially fine saws, individual sheets of plywood are obtained by unrolling a tree on a special peeling machine. After a barked log has been elaborately softened in hot water, it is held in the machine by spikes at either end and turned against a long blade parallel to it. This can be adjusted according to the thickness required, which can be as little as one-hundredth of an inch.

The machine peels a log more skilfully and evenly than one can peel an apple and a slicer cuts the emerging veneers into convenient lengths. After being dried in special kilns, the sheets are

coated with glue, arranged together in whatever number the particular strength demands and then compacted together in a hydraulic hot press.

Plywood used to be associated chiefly with humdrum objects such as tea-chests. Nowadays the Canadians, for example, have developed beautiful pre-finished wall-panelling in hardwood plywood, while Denmark has produced a highly useful fire-resistant compact wood for housing and shipbuilding.

The value of plywood has been immensely increased by powerful modern adhesives or glues, which are helping, too, in the increasing use of laminated timber. Lamination simply means splitting into layers and these layers, when properly bonded together, provide considerable strength, particularly in comparison with cruder methods of joining wood.

Lamination is used, to take one instance, in end-jointing, which is always extremely important from a point of view of strength. One method is to join beams by what is called finger-jointing in which several matching zig-zag cuts are made in each respective end and interleaved into each other. In general, lamination in conjunction with new adhesives and bonding processes, can now provide a greater strength than that of single timbers.

A simpler way of strengthening wood is bending instead of machining it. Properly manufactured, 'bent' articles such as tennis-racket frames and hockey sticks are very much stronger than similar shapes cut out of wood and on this principle wood-bending is being applied on a much bigger scale.

But the most modern method of strengthening timber is by impregnating it with plastics. Timber and plastics are sometimes looked upon as competitors, but considerable thought is being given to their use in co-operation with each other. For each possesses complementary qualities.

43 *Timbine plywood chassis used in the 1800 cc Marcos sports car*

Plastics are waterproof and durable – but expensive and comparatively brittle. Timber is strong and fairly cheap, but susceptible to humidity and decay, against which it has to be protected by expensive paint. Clearly then, if these two materials can be brought together, great advantages will result.

Already aircraft propellers are being made from this plastic-impregnated wood, while once the question of expense has been overcome, it is likely that this new technique of 'wood-plastic composite' or WPC will be applied on a widening scale.

Together with new methods of strengthening timber, the faster and easier use of timber itself is an important new development in human economy. The most spectacular instance of this is in the increased use of timber-frame construction in housing, a process largely developed by Canada and well established now both in North America and Europe, including Britain.

Foundation framing, floor framing, wall framing and roof framing are all constructed in separate units in the factory. They are then brought to the site and erected, after which brickwork and tiling is carried out as necessary. The use of timber-frames is likely to expand because it is economical both in money and man-hours.

Like so many things, timber-framing is an example of ancient methods being revived and enlarged. It is still possible to see in the English countryside houses, three or four hundred years old, that were built on the same principle.

But it is not merely strength or faster use that has to be considered. It is not enough to grow more forests if the timber they produce is not fully exploited. The world need for wood is increasing rapidly and the FAO estimates that by 1975 we shall require nearly 50% more wood and wood products than in 1961. Fifteen hundred million cubic metres of industrial wood will be

44 *Building under construction using the timber-frame method*

needed, quite apart from some 1,200 cubic metres of fuelwood.

Yet in face of this urgent need, three-quarters of every tree felled for timber is wasted. This waste takes many forms – stumps, branches, tops of trees are left in the woods to rot or be burned. Even out of the remaining 25% of the timber that reaches the sawmills, another 10% is wasted in sawdust. In Britain alone it is estimated that every year £45 million worth of timber is lost as more or less useless sawdust.

One new way of using wood waste has been found in the manufacture of synthetic logs by the American 'Pres-to-Log' method. In this, handy, cylindrical eight-pound briquettes are made by crushing the waste, feeding it into a special die and compressing it under 165,000 lbs pressure. These new-style logs are being used extensively for fireplace fuel but could also be burned in railway engines and ships.

So alongside the efforts of the foresters, the wood-technologists are constantly engaged in research into better ways of using wood and its by-products, better ways of protecting it from fire, fungus, insects or weather. All the major timber-producing and timber-using countries maintain research institutes. In Britain the chief agency is the Forest Products Research Institute, an offshoot of the Ministry of Technology, while commercially the Timber Research and Development Association is supported by private building companies and architects. In the United States and Canada, Forest Products Laboratories are supported by the State.

The work of the researchers covers every conceivable aspect of timber usage. New kiln methods for artificially seasoning wood, elaborate 'sun-followers' for exposing experimental materials to maximum exposure, radio-isotype techniques for determining the density of timbers, pressure-impregnation of wood with special preservatives.

A scientist in Sydney Museum carries out laboratory tests
the leaves of eucalyptus trees

Entomologists, botanists, mycologists, chemists, physicists, engineers, electronic experts, fire-prevention specialists, building consultants, and many other technologists and scientists, all combine together. They are determined that the utmost efforts shall be made to utilize fully every aspect of the tree and its constituents.

26 · Last Leaves

This book is incomplete. Within its particular framework it could hardly be otherwise. A separate book could be written on any single aspect it has touched on. But the aim here has been to give some idea of the vast scope and complexity of forestry and timber-usage all over the world, from the days when men believed their dead ancestors dwelt in the trees, to modern times when science is being applied to the subject instead of superstition.

As with all other aspects of Nature, men have slowly come to understand that trees and forests are essential to them both aesthetically and materially. We need them economically to help improve our standard of life, we need them for recreation and to help conserve our threatened wildlife and protect the soil, we need them in our cities to help keep us in touch with the natural world.

All over the world there are organizations working to instil an appreciation of the value of trees in our lives and to encourage the planting of them. The Forestry Commission in Britain maintains intimate relations with many schools, to help them not only in studying the natural history of the tree and the identification of species, but also in planting school forests – an exciting new project. Every year it holds residential forestry courses for schoolteachers, while for would-be professional foresters it conducts a training course lasting about four years, of which two years are spent in a training forest and the remainder at a Forestry Training School.

A society such as 'The Men of the Trees', founded by Richard St Barbe Baker, exists to arouse public opinion to the urgent need to preserve and extend the forests of the world and has branches in many countries. Similarly, the Royal Forestry Society, established nearly a century ago, works particularly for the encouragement of good management of private woodland. It, too, cooperates with schools and students and holds annual examinations for the Woodman's and Forester's Certificates, after which a National Diploma of Forestry can be obtained.

Activities such as these are increasing. Now, after years of destruction, Man is at last beginning to recognize the need to preserve and improve trees, and to realize something of their value.

Index

Figures in bold type refer to the page numbers of illustrations